有机化学：
100个必备反应机理

Organic Chemistry:
100 Must-Know Mechanisms

（美）R. A. 瓦利乌林 编
Roman A. Valiulin

白仁仁 译

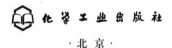
·北京·

内容简介

掌握有机化学的经典反应机理是学好有机化学的关键，也是化学、化工和药物化学科研人员从事专业研究的基础。本书重点介绍了100种经典反应的详细机理。每个机理都以一个完整的信息图示和机理图示进行展示和介绍，并提供了关键概念、规则、缩写和术语的精练信息。有别于同类书，本书的每个反应的机理都以清晰、简明和精美的彩色图示来表述，不仅浅显易懂，而且赏心悦目。

本书适用于从事化学、化工和药物研发等领域的科研院人员及院校师生。

Organic Chemistry: 100 Must-Know Mechanisms /by Roman A. Valiulin
ISBN 978-3-11-060830-4.
Copyright© 2020 by Walter de Gruyter GmbH. All rights reserved.
Authorized translation from the English language edition published by Walter de Gruyter GmbH

本书中文简体字版由Walter de Gruyter GmbH授权化学工业出版社独家出版发行。
本书仅限在中国内地(大陆)销售，不得销往中国香港、澳门和台湾地区。未经许可，不得以任何方式复制或抄袭本书的任何部分，违者必究。

北京市版权局著作权合同登记号：01-2022-5787

图书在版编目（CIP）数据

有机化学：100个必备反应机理/（美）R. A. 瓦利乌林（Roman A. Valiulin）编；白仁仁译. —北京：化学工业出版社，2022.12（2025.1重印）

书名原文：Organic Chemistry: 100 Must-Know Mechanisms
ISBN 978-7-122-42301-6

Ⅰ.①有… Ⅱ.①R…②白… Ⅲ.①有机化学 Ⅳ.①O62

中国版本图书馆CIP数据核字（2022）第180722号

责任编辑：杨燕玲　　　　　　　　　　　　文字编辑：朱　允
责任校对：边　涛　　　　　　　　　　　　装帧设计：史利平

出版发行：化学工业出版社（北京市东城区青年湖南街13号　邮政编码100011）
印　　装：北京缤索印刷有限公司
880mm×1230mm　1/32　印张8　字数191千字　2025年1月北京第1版第3次印刷

购书咨询：010-64518888　　　　　　　　　售后服务：010-64518899
网　　址：http://www.cip.com.cn
凡购买本书，如有缺损质量问题，本社销售中心负责调换。

定　　价：59.80元　　　　　　　　　　　　　　版权所有　违者必究

序言与概述

学习原理。任何一个全新的知识体系对学习者而言似乎都有无限的复杂性，并可能给人难以理解的最初印象，让人心生畏惧。对于有机化学来说更是如此。有机化学这门学科内容繁多，概念复杂且抽象，尤其对于初学者来说，可能会对它更加望而生畏。培养对有机化学的良好直觉，需要我们认真理解、仔细研究。与大多数学科一样，持续的学习和实践可以揭示其内在的模式、共性、规则和逻辑。随着我们成长为更有经验的化学科研人员，有机化学的知识架构也会变得更加清晰。学习有机化学，无疑需要投入一定的时间和精力，且应有持之以恒的努力。要想更好地掌握这一学科，必须重点关注有机反应的机理，并将其作为有机化学学习的基础和支撑点。有机反应的机理非常实用，不仅可以从逻辑上解释分子中的化学键是如何形成或断裂的，而且有助于分析目标产物和副产物形成的原因。此外，随着越来越多的反应机理被解析和掌握，我们将逐渐理清其相似之处，然后在脑海中绘制出一条之前不明显的概念"线索"。这将有助于对相关概念的思考和理解，如反应中间体、过渡态、电荷、自由基和机理箭头等。由此可知，对反应机理的准确理解可为学习有机化学打下坚实的基础，结合实验实操，可将所学知识转化为技能并应用于实践，以更好地服务于科学研究。

编著原则。为了帮助有机化学研究人员提升和"催化"其归纳能力，并为相关研究提供"首选"参考，本书力求对最重要的机理进行简要总结。这也正体现在本书书名之中：《有机化学：100个**必备**反应机理》。本书作者利用本科和研究生阶段学习的知识体系，结合博士后期间有机合成领域的研究工作，完成了本书的编写。基于对增量学习过程的敏锐认识，本书按类别归纳和介绍了相关反应机理，以基础知识和基本机理（如亲核取代及消除），以及最知名的人名反应机理 [如狄尔斯 - 阿尔德环加成（Diels-Alder cycloaddition）、光延反应（Mitsunobu reaction）] 为开端。此外，还补充了有机化学史上重要的机理 [如重氮化反应（diazotization）、卤仿反应（haloform reaction）]。最后，书中还囊括了一些作者心目中非常"精巧"和"酷炫"的机理[如帕特罗-布奇反应（Paternò-Büchi reaction）、炔烃点击化学反应]。

组织框架。本书旨在成为有别于传统教科书的参考指南。为了便于读者参考，在6个基本机理之后，相关机理按照反应名称英文字母排顺，编号从7至100。每个机理都以一个完整的信息图示和机理图示进行展示和介绍，并提供围绕关键概念、规则、缩写和术语的精练信息。作者聚焦于最基本的核心知识，以便于读者的理解和记忆。相信勤奋的读者在始终如一地完成每一个机理的学习、练习、记忆和思考之后会大有收获。本书既可以作为学习起点服务于未来的研究，也可作为随身的工具书。这100种经典、知名的反应机理既是有机化学的理论核心和基础，同时也在实践中得到了广泛应用。当然，本书绝非一本教条式排序或综合的选集。书中反应机理的选择有一定程度作者的主观性，本书内容最初是每日发布于社交媒体上带有标签#100Must-Know Mechanisms 的信息，得到了世界各地学生和化学研究人员的大力支持，作者对相关机理进行了系统的组织和汇编，并增加了额外的历史背景和概念空间，使读者可以方便地学习。

全彩图示。书中每个关键反应都围以蓝框；每个关键机理（本书核心内容）都围以红框；与100种核心机理相关的其他反应和机理都围以灰框；其他实用的规则、概念和信息则围以绿框呈现。书中还包含了介绍反应多样性的图示，脚注中显示了相关的注释和说明。

信息来源。本书的基本信息与经典和主要的有机化学教科书[1]中所包含的信息非常接近。某些反应的信息来源于更专业的文献（如有机金属或光化学转化）[2]。当然，也鼓励读者阅读其他相关的参考书目[3]。由于本书力求展示精简的机理图，因此也鼓励读者参考其他更全面的书籍，特别是关于有机化学人名反应的书籍[4]。此外，精选的开放性在线资源同样实用[5]。当书中呈现的信息准确、全面，且得到参考文献的支持时，将会介绍信息的来源。另外，作者还基于本科、研究生和博士后研究所学习的综合知识进一步对相关信息进行了补充。作者认为《有机试剂百科全书》[6]是个人学习和职业生涯中非常实用的"入门"起点，尤其是在接触新的化学领域或使用新的试剂时具有较高的参考价值。本书中每个机理图示都是以首次提及相关反应或机理的可能原始文献为参考（参见每一机理后的时间轴）。同时，本书还根据需要引用了关键性和基础性综述，阐明机理相关的近期文献以及其他研究文献。本书对机理的总结都是基于作者的合理判断，书中的全部内容经过仔细检查，并且符合行业标准和公认的化学规则，不过仍可能存在个别不妥。需要注意的是，由于本书不是对1800—2020年有机化学反应机理的世界性历史回顾，文本和学术参考文献的多样性并不意味着按时间顺序对文献和历史进行全面介绍。在本书的编写和相关文献的修订过程中，机理本身及人们对机理的理解也在发生变化。因此，读者学习时应该将本书与经验丰富的教授、专家编写的其他综合性教科书、原始文献结合使用。

其他注意要点。同样重要的是，读者需要保持学习的灵活性，并清楚相关机理只是基于我们当前的理解来解释的，同时结合了对基本化学规则、化合价、电子移动规则、电荷守恒、路易斯电子点结构等方面的考量。这些机理可能并不是最"尖端"或最新颖的（如可能无法很好地理解交叉偶联反应），机理也可能是底物依赖性的，并且每个反应可能经历了略有不同的途径。因此，读者不应将本书视为教条式的指南，而应对新数据和创造力保持开放的态度，并将本书视为该领域持续讨论的一部分。

基础知识。要想充分从本书中受益，读者应具备有机化学的基本背景知识。图示都是建立在读者对常见术语和符号的理解基础之上，因此本书中未对基本概念进行介绍或解释。本书适用于有机化学领域的本科生、研究生、教师和科研人员。对于经验丰富的研究人员，如果希望重新回顾和学习有机化学中最基本、最核心的反应机理，本书也不失为一本很好的浓缩"复习"资料。

灵感和深入学习。作者大量借鉴了其从化学专业学生到学术研究员期间积累的个人经验。在学生时期，作者从未学习过关于有机化学反应机理的正式课程，最初仅是通过记忆而不是推导来接触相关知识，第一印象自然是畏惧和不知所措。然而，经过多年的经验积累，作者逐渐掌握了较强的学习和分析反应背后机理的归纳能力，这样的个人经历毫无疑问确立了本书的学习理念，并且本书还将通过视觉和空间有效地传达相关信息。此外，由于大多数人都更有视觉学习的倾向，因此本书更加直观地总结了最实用、最基本的反应机理，很好地契合了这一最便捷的学习模式。从最初的社交媒体开始，经过广泛的讨论、参与，结合对文稿更为细致地准备、组织和策划，作者最终编著成本书。作者希望本书能为需要更多指导和鼓励，正处于成长中的化学研究人员提供一个良好的起点。本书无疑也将激发建设性的讨论，鼓励读者在学习中批判性地思考，去挑战和寻找不同的答案，最终成长为一名化学专家，并保持对这一专业的敏锐直觉。最后，知识是一个类似分形的概念，看得越仔细，发现和掌握的细节就越多。为读者呈现出合理、精准和综合的知识体系，也正是编纂本书的目的。进一步学习的核心读物[1]、原始文献和二次文献[2-4]、在线资源[5,6]，以及实际操作实验和参与实践都将有助于描绘出机理全貌，帮助有机化学科研人员成为见多识广的化学专家。

译者序

掌握有机化学的经典反应机理是学好有机化学的关键，也是化学、化工和药物化学科研人员从事专业研究的基础。反应机理也是我大学阶段有机化学学习的重要组成部分。记得当时我将教材中重要的人名反应和机理归纳整理好，不时翻阅以牢记于心。毕业后，我虽然继续从事药物化学方面的科研工作，在合成目标化合物过程中也时常会用到某一人名反应，但对于不常应用的人名反应，印象似乎渐渐模糊了，只是记得反应的基本定义说明，而具体的反应机理则不一定说得清楚。所以，我便计划将重要的人名反应及其机理再时不时回顾和复习一遍。虽然目前有关人名反应的参考书较多，但重点突出、简洁明了的参考资料却并不多见。一次机缘巧合之下，我发现了由罗曼 A. 瓦利乌林（Roman A. Valiulin）博士于2020年出版的《有机化学：100个必备反应机理》（*Organic Chemistry：100 Must-Know Mechanisms*）一书。阅读几页后，便被该书独特的编排和机理表述方式深深吸引。书中的机理不是追求全而广，而是重点详细介绍了100种经典反应机理。重要的是，每个反应的机理都以清晰、简明和精美的彩色图示来表述，不仅浅显易懂，而且赏心悦目。和传统黑白的图示相比，让人更有兴趣一页页地学习翻阅。与身边的同事及学生分享后，该书也都获得了一致的赞美。

清代诗人袁枚曾说，"书非借不能读也"；一位知名收藏家曾说过，"书买来即便不读，也会给人心灵上的慰藉"。想想自己也常是发现了好书就迫不及待地买来，翻阅几次后便整齐分类置于书架之上，果真有了自己已经读过了的"慰藉"。就这本机理参考书而言，和自己以往翻译的动辄数十万甚至上百万字的药学专著相比，内容相对较少，于是我便决定将其翻译成中文，一方面可督促自己将其从头至尾认真学习和研究，另一方面也可惠及更多的读者。于是利用闲暇时间，完成了本书的翻译和校对。

在阅读翻译本书时，发现原著略显不足之处在于其直接切入主题地介绍机理，而忽略了对人名反应本身的描述和介绍。原著作者曾提及本书的读者须具备一定的有机化学基础，或许这也正是其省略了相关介绍的原因。但阅读过程中如还需时不时翻阅其他资料总归有所不便，因此在翻译中我便以"译者注"的形式加入了对反应本身的概括性介绍，以方便读者学习。同时，在翻译中还改正了原著的一些在细节上的错误。

我的研究生葛嘉敏、钟智超、朱俊龙、赵瑞、李俊杰和蔡红等同学也参与到本书的翻译和校对，在此表示感谢。同时感谢浙江工业大学姜昕鹏教授给予的帮助！

感谢化学工业出版社编辑团队在译书出版过程中的辛勤付出！

尽管译者尽了自己最大的努力，但难免存在不足之处，敬请读者海涵！

白仁仁
renrenbai@126.com
2023年4月于杭州

缩略语表

≡	identical to	等同于
1°	primary [e.g., carbocation] or first generation [e.g., catalyst]	伯、一级［如碳正离子］，或第一代［如催化剂］
2°	secondary [e.g., carbocation] or second generation [e.g., catalyst]	仲、二级［如碳正离子］，或第二代［如催化剂］
3°	tertiary [e.g., carbocation] or third generation [e.g., catalyst]	叔、三级［如碳正离子］，或第三代［如催化剂］
Ac	acetyl	乙酰基
acac	acetylacetonate	乙酰丙酮
Ad_E2	bimolecular electrophilic addition	双分子亲电加成
Ad_E3	trimolecular electrophilic addition	三分子亲电加成
ADMET	acyclic diene metathesis [polymerization]	非环二烯复分解［聚合］
AIBN	azobisisobutyronitrile；2,2′-azobis (2-methylpropionitrile)	偶氮二异丁腈；2,2′-偶氮二（2-甲基丙腈）
Alk=R	alkyl group	烷基
anti	from opposite sides (in anti-addition or anti-elimination)	从对面（反式加成或反式消除）
APA	3-aminopropylamine；1,3-diaminopropane	3-氨基丙胺；1,3-二氨基丙烷
aq	aqueous [work-up]	水性的［后处理］
Ar	aryl；aromatic ring	芳基；芳香环
$B(B^-)$	Brønsted–Lowry base (proton acceptor)	布朗斯特-劳里碱（质子受体）
B_2pin_2	bis (pinacolato) diboron；4,4,4′,4′,5,5,5′,5′-octamethyl-2,2′-bi-1,3,2-dioxaborolane	双（频哪醇合）二硼；4,4,4′,4′,5,5,5′,5′-八甲基-2,2′-联-1,3,2-二氧杂环戊硼烷
9-BBN	9-borabicyclo [3.3.1] nonane	9-硼双环［3.3.1］壬烷
$BH(BH^+)$	Brønsted–Lowry acid (proton donor)	布朗斯特-劳里酸（质子供体）
Bn	benzyl	苄基
Boc	*tert*-butoxycarbonyl；*t*-butoxycarbonyl	叔丁氧羰基

续表

Bs	brosyl; 4-bromobenzenesulfonyl	对溴苯磺酰基;4-溴苯磺酰基
Bu	butyl (if not specified =n-Bu)	丁基(如未指定=n-Bu)
CHD	1,4-cyclohexadiene	1,4-环己二烯
CM=XMET	[olefin] cross-metathesis	烯烃交叉复分解
con	conrotatory	顺旋
3-CR(MCR)	3-component reaction (multi-component reaction)	三组分反应(多组分反应)
4-CR(MCR)	4-component reaction (multi-component reaction)	四组分反应(多组分反应)
CuAAC	copper(Ⅰ)-catalyzed azide-alkyne cycloaddition	铜(Ⅰ)催化的叠氮-炔环加成
CuTC	copper(Ⅰ) thiophene-2-carboxylate	噻吩-2-甲酸亚铜(Ⅰ)
Cy	cyclohexyl	环己基
Cy$_2$BH	dicyclohexylborane	二环己基硼烷
DABCO	1,4-diazabicyclo[2.2.2]octane	1,4-二氮杂双环[2.2.2]辛烷
DBU	1,8-diazabicyclo[5.4.0]undec-7-ene	1,8-二氮杂双环[5.4.0]十一碳-7-烯
DCC	N,N'-dicyclohexylcarbodiimide; 1,3-dicyclohexylcarbodiimide	N,N'-二环己基碳二亚胺;1,3-二环己基碳二亚胺
DCM	dichloromethane; methylene chloride	二氯甲烷
DEAD	diethyl azodicarboxylate	偶氮二甲酸二乙酯
DIAD	diisopropyl azodicarboxylate	偶氮二甲酸二异丙酯
DIBAL= DIBAL-H	diisobutylaluminum hydride=(i-Bu)$_2$AlH	二异丁基氢化铝
dis	disrotatory	对旋
DMAP	4-dimethylaminopyridine; 4-(dimethylamino)pyridine	4-二甲氨基吡啶;4-(二甲氨基)吡啶
DMP	Dess-Martin periodinane	戴斯-马丁高碘试剂
DMSO	dimethyl sulfoxide	二甲基亚砜
E-	entgegen (*trans*-or opposite)	异侧(反式或相反)
e$^-$	electron	电子

E（或 E⁺）	electrophile	亲电体（亲电试剂）
E1	unimolecular elimination	单分子消除
E1cB(E1cb)	unimolecular elimination conjugate base	单分子消除共轭碱
E2	bimolecular elimination	双分子消除
EDC=EDCI	l-ethyl-3-(3′-dimethylaminopropyl) carbodiimide hydrochloride；N-(3-dimethylaminopropyl)-N′-ethylcarbodiimide hydrochloride	1-乙基-3-（3′-二甲基氨基丙基）碳二亚胺盐酸盐；N-（3-二甲氨基丙基）-N′-乙基碳二亚胺盐酸盐
EDG=ERG	electron donating group	给（供）电子基团
E_i	internal or intramolecular elimination	分子内消除
eq	equivalent	当量
ERG=EDG	electron releasing group	电子释放基团（与 EDG 相同）
Et_2BH	diethylborane	二乙基硼烷
EWG	electron withdrawing group	吸电子基团
EYM	enyne metathesis	烯炔复分解
Grubbs1°	the Grubbs catalyst first generation	第一代格拉布催化剂
Grubbs 2°	the Grubbs catalyst second generation	第二代格拉布催化剂
$H_3B·THF$	borane-tetrahydrofuran complex；borane tetrahydrofuran complex	硼烷-四氢呋喃复合物
$H_3B·Me_2S$=BMS	borane-dimethyl sulfide complex；borane dimethyl sulfide complex	硼烷-二甲硫醚复合物
HATU	N-[(dimethylamino)-1H-1,2,3-triazolo[4,5-b]pyridin-1-ylmethylene]-N-methylmethanaminium hexafluorophosphate N-oxide; 1-[bis(dimethylamino)methylene]-1H-1,2,3-triazolo[4,5-b]pyridinium 3-oxide hexafluorophosphate	2-（7-偶氮苯并三氮唑）-N,N,N′,N′-四甲基脲六氟磷酸酯
HBTU	O-benzotriazol-1-yl-N,N,N′,N′-tetramethyluronium hexafluorophosphate; 3-[bis(dimethylamino)methyliumyl]-3H-benzotriazol-1-oxide hexafluorophosphate	O-苯并三氮唑-N,N,N′,N′-四甲基脲六氟磷酸酯
HET=^{HET}Ar	heterocycle；heteroaromatic ring；heteroaryl	杂环；芳杂环；芳杂基

续表

HOAt=HOAT	1-hydroxy-7-azabenzotriazole；3-hydroxy-3H-1,2,3-triazolo [4,5-b] pyridine	1-羟基-7-氮杂苯并三氮唑；3-羟基-3H-1,2,3-三氮唑并［4,5-b］吡啶
HOBt=HOBT	1-hydroxybenzotriazole	1-羟基苯并三唑
HOMO	highest occupied molecular orbital	最高占据分子轨道
$h\nu$	light (direct irradiation) or excited state	光（直接照射）或激发态
$I_i(BR)$	intermediate (biradical)	中间体（双自由基）
$I_i(RP)$	intermediate (radical pair)	中间体（自由基对）
IBX	2-iodoxybenzoic acid；o-iodoxybenzoic acid	2-碘氧基苯甲酸；邻亚碘酰基苯甲酸
IC	internal conversion	分子内转换
Ipc_2BH	diisopinocampheylborane	二异松莰基硼烷
$IpcBH_2$	monoisopinocampheylborane	单异松莰基硼烷
ISC	intersystem crossing	系间穿越
KAPA	potassium 3-aminopropylamide	3-氨基丙胺钾
L	ligand or leaving group	配体或离去基团
(l)	liquid [as in liquid ammonia：NH_3 (l)]	液体[如液氨：NH_3（l）]
LA	Lewis acid	路易斯酸
LAPA	lithium 3-aminopropylamide	3-氨基丙胺锂
LDA	lithium diisopropylamide=$(i\text{-}Pr)_2NLi$	二异丙基氨基锂
L_mPd	palladium (0) cross coupling catalyst	钯(0)交叉偶联催化剂
L_nPd	low-coordinate palladium (0) cross coupling catalyst	低配位钯(0)交叉偶联催化剂
LUMO	lowest unoccupied molecular orbital	最低未占分子轨道
M	metal	金属
[M]	metal catalyst (not specified)	金属催化剂（如未指定）
$M^{+3}=M(Ⅲ)$	oxidation state (oxidation number) of an element [e.g., $Cu^{2+}=Cu(Ⅱ)$；$Pd^0=Pd(0)$]	元素的氧化态（氧化数）[如 $Cu^{2+}=Cu(Ⅱ)$；$Pd^0=Pd(0)$]
M^{3+}	charge [e.g., Ti^{3+} in $TiCl_3$ versus $Ti^{+3}=Ti(Ⅲ)$]	电荷[如 $TiCl_3$ 中的 Ti^{3+} 与 $Ti^{+3}=Ti(Ⅲ)$]

续表

m-CPBA (MCPBA)	*meta*-chloroperbenzoic acid; *m*-chloroperbenzoic acid; 3-chloroperbenzoic acid	间氯过氧苯甲酸; 3-氯过氧苯甲酸
MCR	multi-component reaction	多组分反应
Mes	mesityl (from mesitylene=1,3,5-trimethylbenzene)	均三甲苯=1,3,5-三甲基苯
Ms	mesyl; methanesulfony=SO$_2$Me	甲磺酰基=SO$_2$Me
n	nonbonding [molecular] orbital	非键[分子]轨道
NACM	nitrile-alkyne cross-metathesis	腈-炔交叉复分解
NBS	*N*-bromosuccinimide; 1-bromo-2,5-pyrrolidinedione	*N*-溴代丁二酰亚胺; 1-溴-2,5-吡咯烷二酮
N-HBTU	1-[bis(dimethylamino)methylene]-1*H*-benzotriazolium hexafluorophosphate 3-oxide	1-[双(二甲氨基)亚甲基]-1*H*-苯并三唑六氟磷酸3-氧化物
NiAAC	nickel-catalyzed azide-alkyne cycloaddition	镍催化的叠氮-炔环加成
NMM	*N*-methylmorpholine; 4-methylmorpholine	*N*-甲基吗啉; 4-甲基吗啉
NMO	*N*-methylmorpholine *N*-oxide; 4-methylmorpholine *N*-oxide	*N*-甲基吗啉 *N*-氧化物; 4-甲基吗啉 *N*-氧化物
Ns	nosyl; 2-nitrobenzenesulfonyl	硝基苯磺酰基; 2-硝基苯磺酰基
Nu(或Nu⁻)	nucleophile	亲核体(亲核试剂)
NuH	Brønsted–Lowry acid (proton donor, like BH)	布朗斯特-劳里酸(质子供体, 如BH)
[O]	oxidant (e.g., 2KHSO$_5$·KHSO$_4$·K$_2$SO$_4$)	一般氧化剂(如2KHSO$_5$·KHSO$_4$·K$_2$SO$_4$)
O-HBTU	*N*-[(1*H*-benzotriazol-1-yloxy)(dimethylamino)methylene]-*N*-methylmethanaminium hexafluorophosphate	*N*-[(1*H*-苯并三唑-1-基氧基)(二甲氨基)亚甲基]-*N*-甲基甲胺六氟磷酸盐
p[sp, sp^2, sp^3]	p orbital	p轨道
P	product [in photochemical reactions]	产物[在光化学反应中]
PCC	pyridinium chlorochromate	氯铬酸吡啶盐

PDC	pyridinium dichromate	重铬酸吡啶盐
Ph	phenyl	苯基
Ph₃P=TPP	triphenylphosphine	三苯基膦
PhthNH	phthalimide (Phth=phthaloyl)	邻苯二甲酰亚胺（Phth= 邻苯二甲酰）
pK_a	acidity constant=$-\lg(K_a)$	酸度常数 =$-\lg(K_a)$
Pr	propyl (if not specified =n-Pr)	丙基（如未指定 =n-Pr）
Py	pyridine	吡啶
R	reactant; starting material [in photochemical reactions]	反应物；起始物料［在光化学反应中］
R($-R^1$, $-R^2$, $-R'$, $-R''$, …)	group; alkyl group; substituent; [molecular] fragment	基团；烷基；取代基；[分子] 片段
R*	excited reactant [in photochemical reactions]	激发态反应物［在光化学反应中］
RCAM	ring-closing alkyne metathesis	关环炔烃复分解
RCEYM	ring-closing enyne metathesis	关环烯炔复分解
RCM	ring-closing metathesis	关环复分解
R$_L$	large group (substituent)	大基团（取代基）
ROM	ring-opening metathesis	开环复分解
ROMP	ring-opening metathesis polymerization	开环复分解聚合
R$_s$	small group (substituent)	小基团（取代基）
RuAAC	ruthenium-catalyzed azide-alkyne cycloaddition	钌催化的叠氮 - 炔环加成
s[sp, sp², sp³]	s orbital	s 轨道
S$_0$	ground state	基态
S$_1$	first [energy level] singlet excited state	第一［能级］单重激发态
S$_2$	second [energy level] singlet excited state	第二［能级］单重激发态
S$_E$Ar=S$_E$(Ar)=S$_E$2Ar	[bimolecular] aromatic electrophilic substitution =areniumion mechanism	［双分子］芳香族亲电取代 = 鎓离子机理
³sens	sensitized irradiation [to the triplet excited state]	敏化辐照（至三重激发态）

续表

SET	single electron transfer	单电子转移
Sia$_2$BH	disiamylborane; bis (1,2-dimethylpropyl) borane	二戊基硼烷;双(1,2-二甲基丙基)硼烷
S$_N$1	unimolecular nucleophilic substitution	单分子亲核取代
S$_N$2	bimolecular nucleophilic substitution	双分子亲核取代
S$_N$Ar=S$_N$2Ar	[bimolecular] aromatic nucleophilic substitution	[双分子]芳香亲核取代
S$_{RN}$1	unimolecular radical nucleophilic substitution	单分子自由基亲核取代
syn	from the same side (in syn-addition or syn-elimination)	同侧(在同侧加成或同侧消除)
T$_1$	first [energy level] triplet excited state	第一[能级]三重激发态
T$_2$	second [energy level] triplet excited state	第二[能级]三重激发态
TBAF	tetrabutylammonium (tetra-n-butylammonium) fluoride =n-Bu$_4$NF	四(正)丁基氟化铵=n-Bu$_4$NF
Tf	triflyl; trifluoromethanesulfonyl=SO$_2$CF$_3$	三氟甲磺酰基=SO$_2$CF$_3$
TFA	trifluoroacetic acid	三氟乙酸
TFAA	trifluoroacetic anhydride	三氟乙酸酐
THF	tetrahydrofuran	四氢呋喃
Thx$_2$BH$_2$	(2-methylpentan-2-yl) borane	(2-甲基戊烷-2-基)硼烷
TLC	thin-layer chromatography	薄层色谱法
TMEDA	N,N,N',N'-tetramethylethylenediamine; 1,2-bis (dimethylamino) ethane	四甲基乙二胺
TMS	trimethylsilyl=SiMe$_3$	三甲基硅烷基
TPAP	tetrapropylammonium (tetra-n-propylammonium) perruthenate= $(n$-Pr$)_4$NRuO$_4$	过钌酸四丙基铵盐
TPP=Ph$_3$P	triphenylphosphine	三苯基膦
Ts	tosyl; p-toluenesulfonyl	甲苯磺酰基;对甲苯磺酰基
X(-X)	halogen or a general leaving group (see L)	卤素或一般离去基团(参见L)

续表

X(=X)	variable atom; variable group (usually O or N)	可变原子或可变基团（通常为 O 或 N）
XMET=CM	[olefin] cross-metathesis	[烯烃]交叉复分解
Z-	*zusammen* (*cis-* or same)	同侧（顺式或相同）
Z(-Z)	variable group (often EWG)	可变基团（通常为 EWG）
α	alpha position (first position)	α 位（1 位）
β	beta position (second position)	β 位（2 位）
γ	gamma position (third position)	γ 位（3 位）
△	temperature; heat or ground state [in photochemical reactions]	温度；加热或基态[在光化学反应中]
δ^+	partial positive charge (low electron density)	部分正电荷（低电子密度）
δ^-	partial negative charge (high electron density)	部分负电荷（高电子密度）
π	involving a π-bond (for example, π-complex)	涉及 π 键（如 π-复合物）
$1\pi\ e^-, 2\pi\ e^-, \ldots$	number of electrons in a π-orbital	π-轨道上的电子数
σ	involving a σ-bond (for example, σ-complex)	涉及 σ 键（如 σ-复合物）
σ^*	[antibonding] sigma star [molecular] orbital	反键[分子]轨道
Φ_{ISC}	quantum yield [for intersystem crossing]	量子产率[用于系间穿越]

目录

反应机理 — 001

1. 亲电加成机理 (electrophilic addition mechanism) — 002
2. 亲核取代机理 (nucleophilic substitution mechanism) — 004
3. 芳香亲电取代机理 (aromatic electrophilic substitution mechanism) — 006
4. 芳香亲核取代机理 (aromatic nucleophilic substitution mechanism) — 008
5. 芳香自由基亲核取代机理 (aromatic radical nucleophilic substitution mechanism) — 010
6. 消除机理 (elimination mechanism) — 014
7. 酮醇缩合 (acyloin condensation) — 018
8. 炔烃拉链反应 (alkyne zipper reaction) — 020
9. 阿尔布佐夫反应 (Arbuzov reaction) — 022
10. 阿恩特－艾斯特尔特合成 (Arndt-Eistert synthesis) — 024
11. 拜耳－维利格氧化 (Baeyer-Villiger oxidation) — 026
12. 巴顿脱羧 (Barton decarboxylation) — 028
13. 贝里斯－希尔曼反应 (Baylis-Hillman reaction) — 030
14. 贝克曼重排 (Beckmann rearrangement) — 032
15. 安息香缩合 (benzoin condensation) — 034
16. 苯炔机理 (benzyne mechanism) — 036
17. 伯格曼环化 (Bergman cyclization) — 038
18. 伯奇还原 (Birch reduction) — 040
19. 比施勒－纳皮耶拉尔斯基环化 (Bischler-Napieralski cyclization) — 042
20. 布朗硼氢化 (Brown hydroboration) — 044
21. 布赫瓦尔德－哈特维希交叉偶联 (Buchwald-Hartwig cross coupling) — 046
22. 坎尼扎罗反应 (Cannizzaro reaction) — 048

23	陈－埃文斯－兰交叉偶联 (Chan-Evans-Lam cross coupling)	050
24	齐齐巴宾氨基化 (Chichibabin amination)	052
25	克莱森缩合 (Claisen condensation)	054
26	克莱森重排 (Claisen rearrangement)	056
27	柯普消除 (Cope elimination)	058
28	柯普重排 (Cope rearrangement)	060
29	克里奇氧化和马拉普拉德氧化 (Criegee & Malaprade oxidation)	062
30	铜催化叠氮－炔环加成 (copper-catalyzed azide-alkyne cycloaddition, CuAAC)	064
31	柯提斯重排 (Curtius rearrangement)	066
32	达森缩合 (Darzens condensation)	070
33	戴斯－马丁氧化 (Dess-Martin oxidation)	072
34	重氮化反应 (diazotization)	074
35	狄尔斯－阿尔德环加成 (Diels-Alder cycloaddition)	076
36	双-π-甲烷重排 (di-π-methane rearrangement)	078
37	法沃尔斯基重排 (Favorskii rearrangement)	080
38	费歇尔吲哚合成 (Fischer indole synthesis)	082
39	傅－克酰基化和傅－克烷基化 (Friedel-Crafts acylation & alkylation)	084
40	盖布瑞尔合成 (Gabriel synthesis)	086
41	格瓦尔德反应 (Gewald reaction)	088
42	格拉泽－埃格林顿－海偶联 (Glaser-Eglinton-Hay coupling)	090
43	格氏反应 (Grignard reaction)	092
44	格罗布裂解 (Grob fragmentation)	094
45	卤仿反应 (haloform reaction)	096
46	赫克交叉偶联 (Heck cross coupling)	098
47	黑尔－福尔哈德－泽林斯基反应 (Hell-Volhard-Zelinsky reaction)	100
48	桧山交叉偶联 (Hiyama cross coupling)	102
49	霍夫曼消除 (Hofmann elimination)	104

50	霍纳－沃兹沃斯－埃蒙斯烯烃化 (Horner-Wadsworth-Emmons olefination)	106
51	琼斯氧化 (Jones oxidation)	108
52	库切罗夫反应 (Kucherov reaction)	110
53	熊田交叉偶联 (Kumada cross coupling)	112
54	莱伊－格里菲斯氧化 (Ley-Griffith oxidation)	114
55	利贝斯金德－斯罗格尔交叉偶联 (Liebeskind-Srogl cross coupling)	116
56	曼尼希反应 (Mannich reaction)	118
57	麦克默里偶联 (McMurry coupling)	120
58	麦尔外因－庞多夫－维利还原 (Meerwein-Ponndorf-Verley reduction)	122
59	迈克尔加成 (Michael addition)	124
60	米尼希反应 (Minisci reaction)	126
61	光延反应 (Mitsunobu reaction)	128
62	宫浦硼化 (Miyaura borylation)	130
63	向山氧化还原水合 (Mukaiyama RedOx hydration)	132
64	纳扎罗夫环化 (Nazarov cyclization)	134
65	尼夫反应 (Nef reaction)	136
66	根岸交叉偶联 (Negishi cross coupling)	138
67	诺里什 I 型和 II 型反应 (Norrish type I & II reaction)	140
68	烯烃复分解 [olefin (alkene) metathesis]	142
69	欧芬脑尔氧化 (Oppenauer oxidation)	146
70	臭氧分解 (ozonolysis)	148
71	帕尔－克诺尔合成 (Paal-Knorr syntheses)	150
72	帕特罗－布奇反应 (Paternò-Büchi reaction)	154
73	保森－坎德反应 (Pauson-Khand reaction)	156
74	肽 (酰胺) 偶联 [peptide (amide) coupling]	158
75	皮克特－斯彭格勒反应 (Pictet-Spengler reaction)	162
76	频哪醇－频哪酮重排 (pinacol-pinacolone rearrangement)	164

77	波罗诺夫斯基反应 (Polonovski reaction)	166
78	普里莱扎耶夫环氧化 (Prilezhaev epoxidation)	168
79	普林斯反应 (Prins reaction)	170
80	普默勒重排 (Pummerer rearrangement)	172
81	兰伯格－贝克伦德重排 (Ramberg-Bäcklund rearrangement)	174
82	雷福尔马茨基反应 (Reformatsky reaction)	176
83	罗宾逊环化 (Robinson annulation)	178
84	夏皮罗反应 (Shapiro reaction)	180
85	薗头交叉偶联反应 (Sonogashira cross coupling)	182
86	施陶丁格反应 (Staudinger reaction)	184
87	斯特格里奇酯化 (Steglich esterification)	186
88	施蒂勒交叉偶联反应 (Stille cross coupling)	188
89	铃木交叉偶联反应 (Suzuki cross coupling)	190
90	斯文氧化 (Swern oxidation)	192
91	乌吉反应 (Ugi reaction)	194
92	乌尔曼芳基－芳基偶联 (Ullmann aryl-aryl coupling)	196
93	厄普约翰双羟基化反应 (Upjohn dihydroxylation)	198
94	维尔斯迈尔－哈克反应 (Vilsmeier-Haack reaction)	200
95	瓦克氧化 (Wacker oxidation)	202
96	瓦格纳－麦尔外因重排 (Wagner-Meerwein rearrangement)	204
97	温勒伯酮合成 (Weinreb ketone synthesis)	206
98	维蒂希反应 (Wittig reaction)	208
99	沃尔－齐格勒反应 (Wohl-Ziegler reaction)	210
100	沃尔夫－凯惜纳还原 (Wolff-Kishner reduction)	212

参考书目与文献	214
中文索引	229
英文索引	233

反应机理

1 亲电加成机理
(electrophilic addition mechanism)

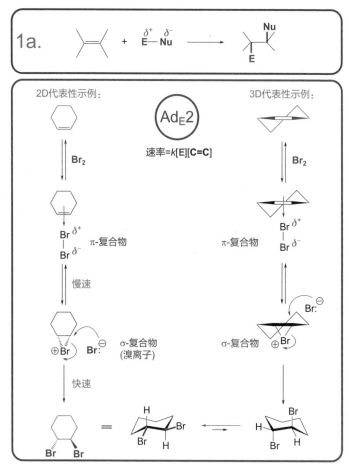

图 1.1 双分子亲电加成（Ad_E2）反应的机理 ❶

---

❶ Ad_E2：双分子亲电加成 [addition electrophilic **bi**-molecular **(2)**]，即反应速率为二级，限速步骤（即慢速步骤）取决于两种反应物的浓度。在环己烯的溴化加成反应中，反应物为亲电试剂（electrophile，E 或 Br_2）和烯烃（C=C），反应速率 $=k[E][C=C]$。

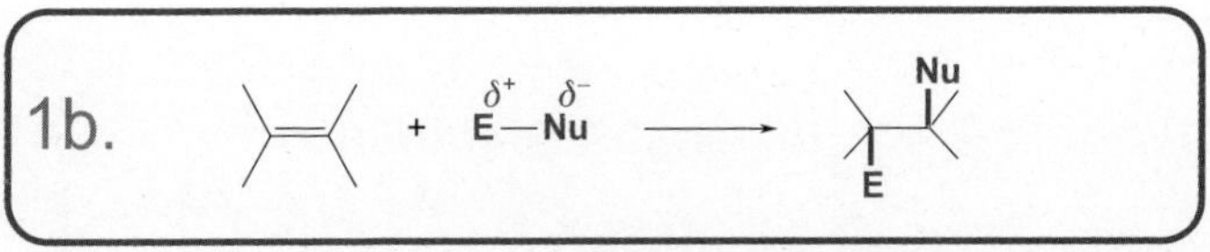

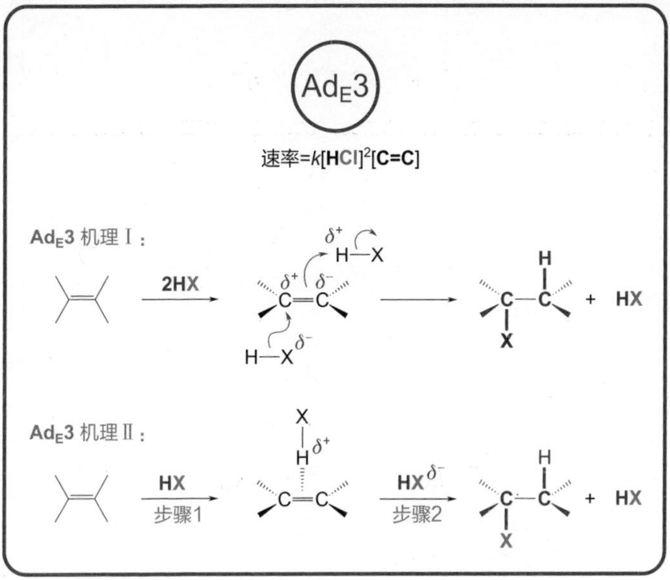

图 1.2 三分子亲电加成（Ad$_E$3）反应的机理 ❶

---

❶ Ad$_E$3：三分子亲电加成 [addition electrophilic **tri**-molecular **(3)**]，即反应速率为三级，限速步骤取决于三种反应物的浓度。在这个不太常见的实例中，反应物为两分子亲电试剂（electrophiles，2HX 或 HCl+HCl）和烯烃（C═C），反应速率 =k[HCl][HCl][C═C] = k[HCl]^2[C═C]。在机理 I 中所有三种组分碰撞的可能性和同时性都比较小。在更可能的机理 II 中，首先第一个 HX 和烯烃之间形成络合物（步骤 1），然后加成第二个 HX（步骤 2）。

2 亲核取代机理
(nucleophilic substitution mechanism)

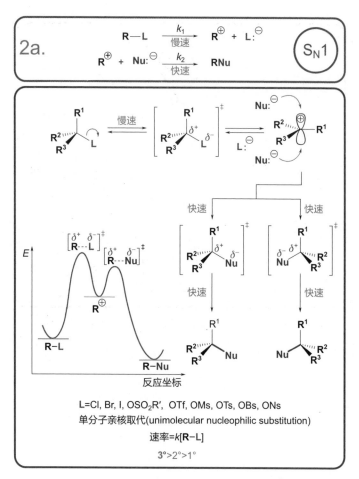

图 2.1 单分子亲核取代（S_N1）反应的机理 ❶

---

❶ S_N1：单分子亲核取代 [substitution **n**ucleophilic **uni**-molecular (**1**)]，即反应速率为一级，限速步骤取决于一种反应物的浓度。在本例中，反应物为含有离去基团（R—L）的起始原料（底物）；反应速率 $=k$[R—L]。

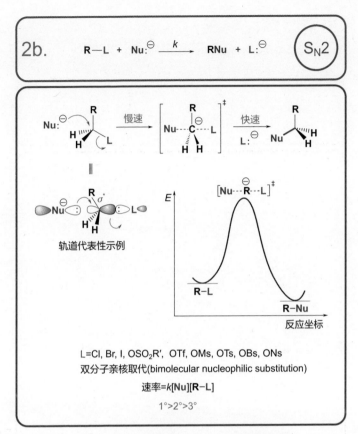

图 2.2 双分子亲核取代（S_N2）反应的机理❶

---

❶ S_N2：双分子亲核取代 [substitution nucleophilic bi-molecular (2)]，即反应速率为二级，限速步骤取决于两种反应物的浓度。在本例中，反应物为亲核试剂（Nu）和起始原料（R—L）；反应速率 $=k[\text{Nu}][\text{R—L}]$。

3 芳香亲电取代机理
(aromatic electrophilic substitution mechanism)

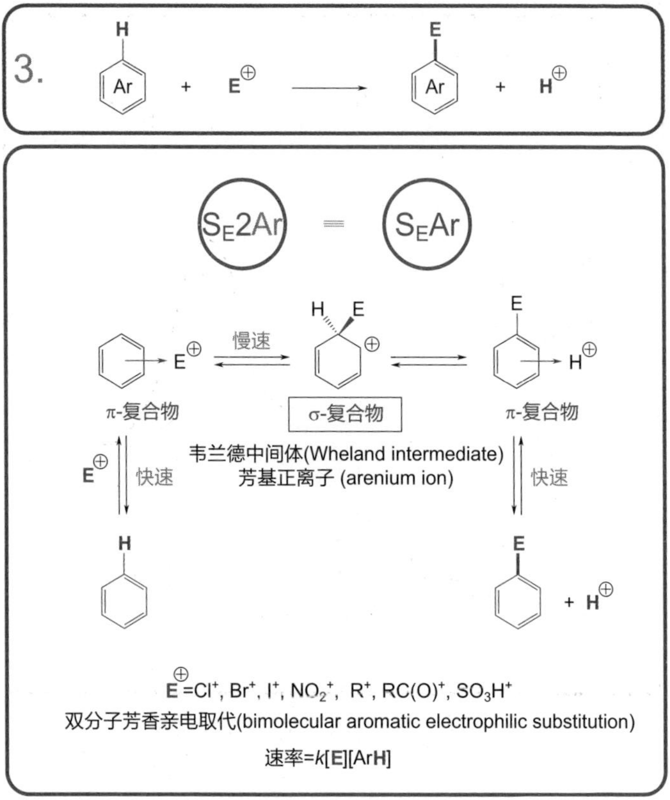

$$3.\quad Ar-H + E^{\oplus} \longrightarrow Ar-E + H^{\oplus}$$

$S_E2Ar = S_EAr$

π-复合物 ⇌(慢速) σ-复合物 ⇌ π-复合物

σ-复合物:韦兰德中间体(Wheland intermediate) 芳基正离子(arenium ion)

$E^{\oplus}$ ↕快速 …… ↕快速

$E^{\oplus}=Cl^+, Br^+, I^+, NO_2^+, R^+, RC(O)^+, SO_3H^+$

双分子芳香亲电取代(bimolecular aromatic electrophilic substitution)

速率 $=k[E][ArH]$

图 3.1 芳基正离子（S_EAr）反应的机理 ❶

---

❶ S_EAr 或 $S_E(Ar)$：芳基正离子亲电取代 [substitution electrophilic arenium (ion)] 机理，即芳基正离子机理。在本例中为双分子(2)反应，即反应速率为二级，限速步骤(即慢速步骤)取决于两种反应物的浓度。反应物为亲电试剂(E)和芳烃(ArH)；反应速率 $=k$ [E][ArH]。为了强调该反应为双分子机理，有时使用 S_E2Ar 或 $S_E2(Ar)$ 表示(简单地使用 S_E2 来表示可能会令人混淆，因为其也适用于脂肪族亲电取代)。

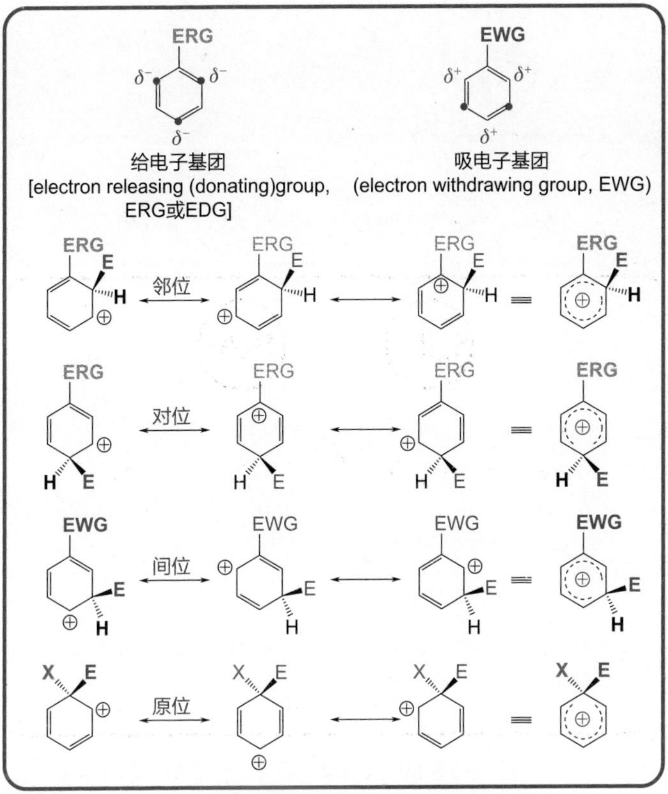

图 3.2 含 EWG 和 ERG 底物的取代定位 ❶

---

❶ 在本书中,给电子基团 ERG 和 EDG 是交替使用的。注意,原位(*ipso-*)取代仅用于与邻位(*ortho-*)、对位(*para-*)和间位(*meta-*)取代的比较。

4 芳香亲核取代机理
(aromatic nucleophilic substitution mechanism)

$$4. \quad Ar\text{-}X + Nu:^{\ominus} \longrightarrow Ar\text{-}Nu + X:^{\ominus}$$

$S_N2Ar = S_NAr$

EWG-ArX + Nu:$^\ominus$ ⇌ (加成, 慢速) [EWG-Ar(X)(Nu)]$^\ominus$ σ-复合物 迈森海默复合物(Meisenheimer complex) → (消除, 快速) EWG-Ar-Nu + X:$^\ominus$

双分子芳香亲核取代(bimolecular aromatic nucleophilic substitution)
(加成-消除)
速率=k[Nu][ArX]

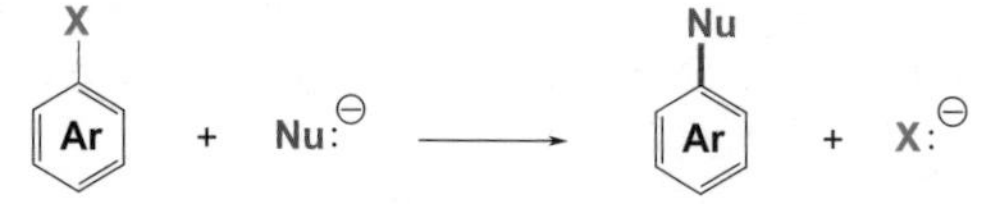

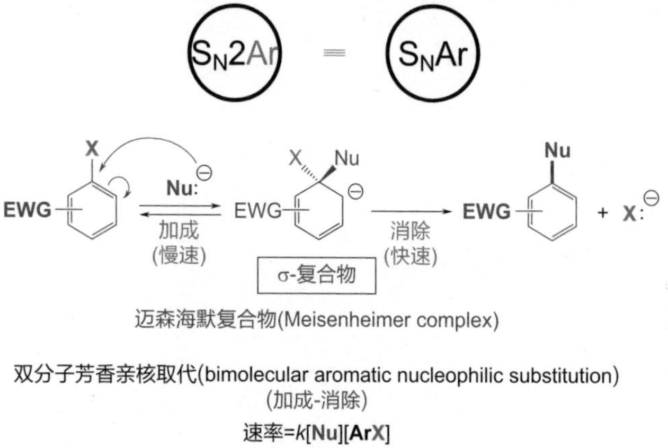

图 4.1 双分子芳香亲核取代(加成-消除)反应的机理(S_NAr)❶

---

❶ S_NAr: 芳香亲核取代(**s**ubstitution **n**ucleophilic **a**romatic)机理,也称为加成-消除机理。在本例中为双分子(2)反应,即反应速率为二级,限速步骤(即慢速步骤)取决于两种反应物的浓度。反应物为亲核试剂(Nu)和芳烃(ArX);反应速率=k[Nu][ArX]。为了强调该反应为双分子机理,有时候使用 S_N2Ar 表示。

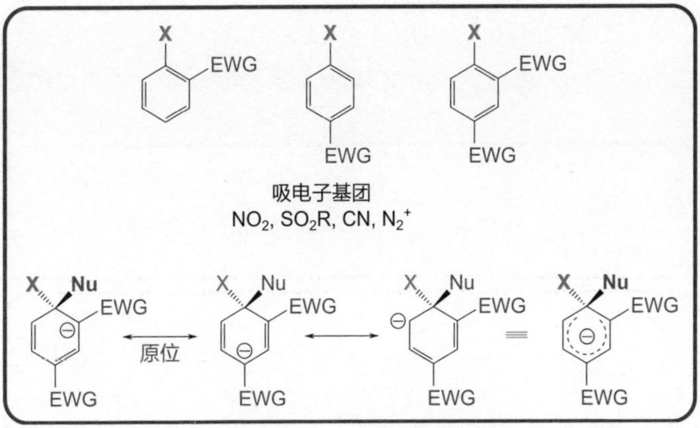

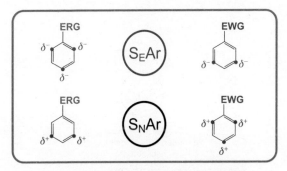

图 4.2　典型活化的 S_NAr 底物 ❶

图 4.3　S_EAr 和 S_NAr 反应中的取代定位 ❷

---

❶ 一个典型的 S_NAr 底物通常包含一个活化的吸电子基团（EWG）和一个离去基团（X）。

❷ 在 S_EAr 反应中，EWG 基团定位间位取代，而 ERG（EDG）基团定位邻位或对位取代。然而，在 S_NAr 反应中，这种基团定位是相反的：EWG 基团定位邻位或对位取代，而 ERG（EDG）定位间位取代。

5 芳香自由基亲核取代机理
(aromatic radical nucleophilic substitution mechanism)

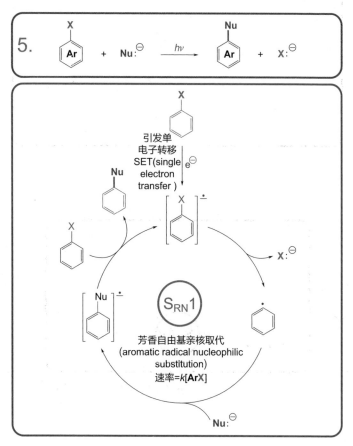

图 5.1 单分子芳香自由基亲核取代（$S_{RN}1$）反应的机理 ❶

---

❶ $S_{RN}1$：单分子（1）自由基亲核取代[**s**ubstitution **r**adical **n**ucleophilic **uni**-molecular **(1)**]，即反应速率为一级，限速步骤（即慢速步骤）取决于一种反应物的浓度。在本例中，反应物为含有离去基团（ArX）的起始原料；反应速率 $=k$ [ArX]。

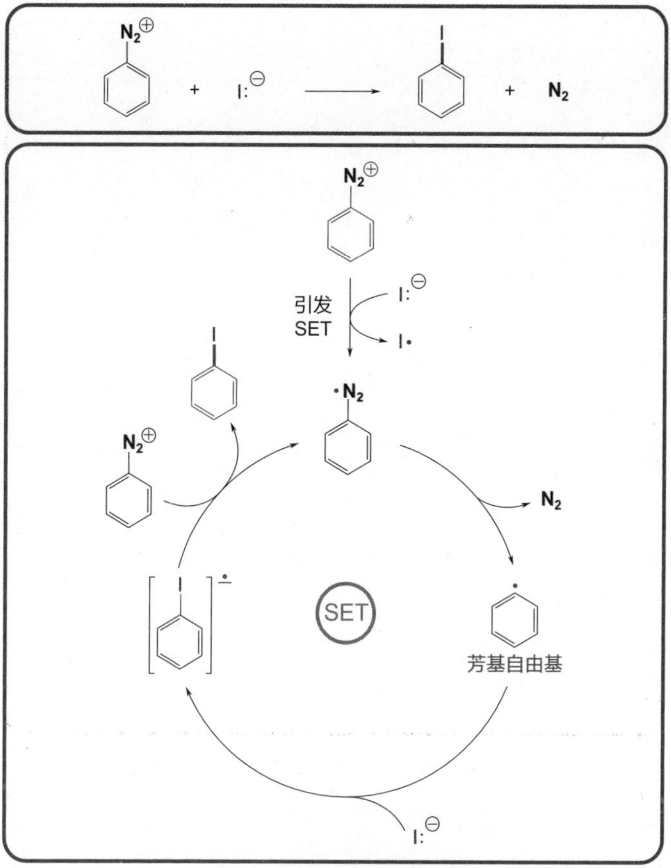

图 5.2 碘对重氮基团的取代 ❶

---

❶ 碘化物取代重氮基团是单电子转移(SET)机理的一个实例。注意：$S_{RN}1$ 机理和 SET 机理是密切相关的，在本书中没有区别。Jerry March 区分了 $S_{RN}1$ 机理(最初为由电子供体对芳基底物的攻击)和 SET 机理(最初为亲核试剂的进攻)。桑德迈尔反应(Sandmeyer reaction)机理与其相关[参见 https://doi.org/10.1002/cber.18840170219 和 https://doi.org/10.1002/cber.188401702202，2019 年 12 月 5 日登录]。

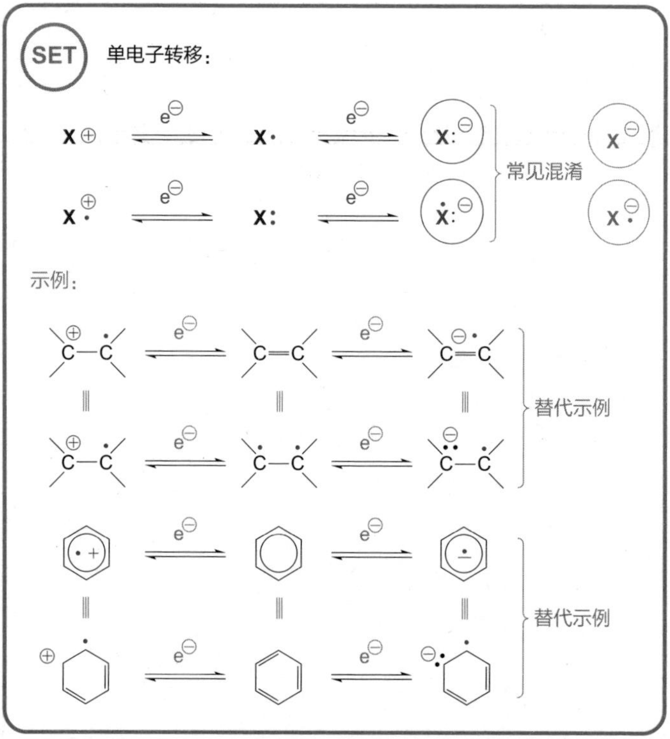

图 5.3　SET 中涉及的自由基路易斯电子点的结构 ❶

---

❶ 本图总结了各种 SET 过程的路易斯电子点结构：阳离子→自由基→阴离子或阳离子 - 自由基→双自由基或孤对电子→阴离子 - 自由基，并显示了几个常见实例。注意：在文献中，阳离子 - 自由基通常被称为自由基阳离子（radical cation），阴离子 - 自由基被称为自由基阴离子（radical anion）。在某些情况下，与阴离子或阴离子 - 自由基相关的孤对电子未清晰表示（有时这种简化会引起混淆）。

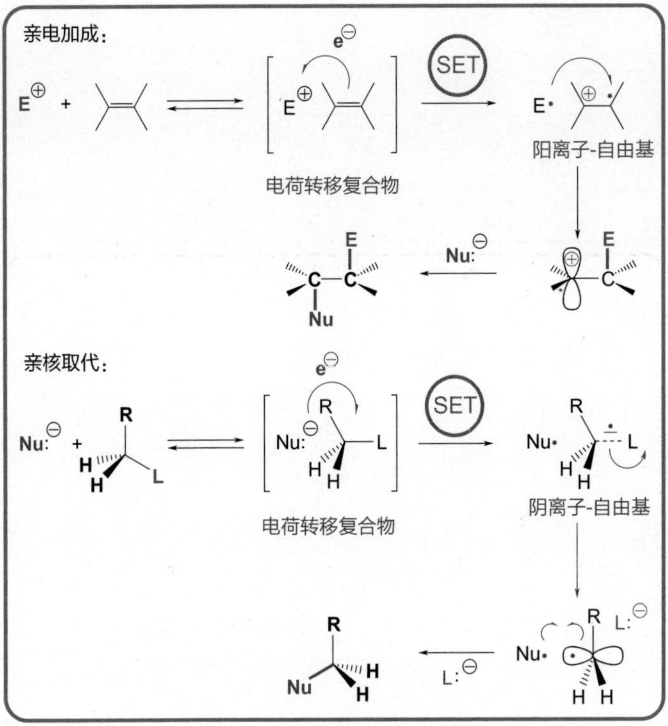

图 5.4　单电子转移机理的实例 ❶

---

❶ 单电子转移（SET）机理描述的亲电加成的一个实例：一个单电子从烯烃转移至亲电试剂，形成一个阳离子-自由基（自由基阳离子）。SET 机理描述的亲核取代的一个实例：一个单电子从亲核试剂转移至底物，形成一个阴离子-自由基（自由基阴离子）[3]。

6 消除机理
(elimination mechanism)

6a. E1cB
速率=$k[\mathbf{B}][RL]$

$\underset{Y\ \ L}{\overset{H\ \ R}{\diagdown\alpha\ \diagup}}\xrightarrow{\text{碱}}\underset{Y}{\overset{R}{\diagdown\diagup}}+HL$

（图示三种情况）

B = 一般碱
Y = EWG等
L = 离去基团

(E1cB)$_R$ = (E1cB)$_{rev}$
A：速率 ≈ $\dfrac{k[\mathbf{B}][RL]}{[BH]}$

(E1cB)$_I$ = (E1cB)$_{irr}$
B：速率 = $k[\mathbf{B}][RL]$

(E1cB)$_{阴离子}$ = (E1)$_{阴离子}$
C：速率 = $k[\mathbf{B}][RL] \approx k[RL]$

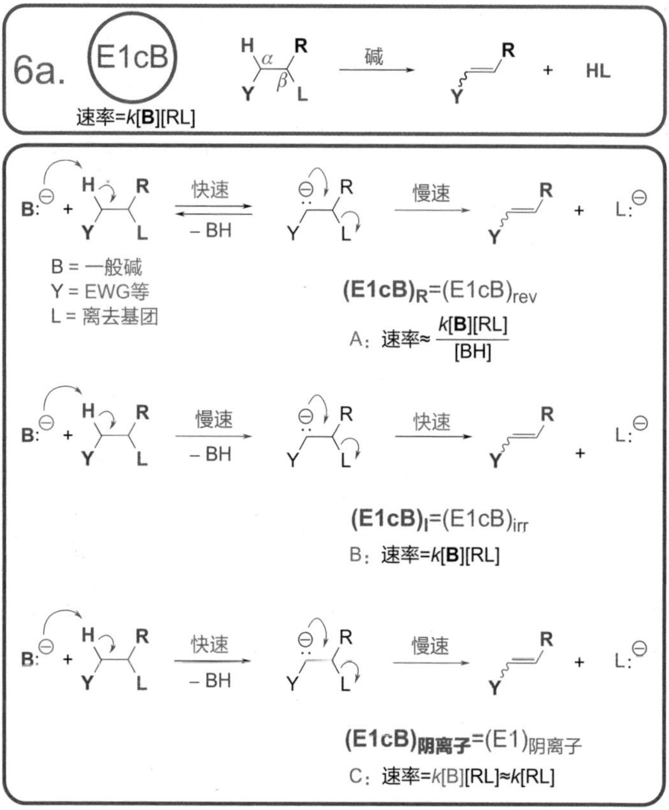

图 6.1 单分子 β- 消除（E1cB）反应的机理 ❶

---

❶ E1cB（E1cb）：单分子（1）共轭碱消除 [elimination **uni**-molecular **(1)** **c**onjugate **b**ase (**b**ase)] 机理，又被称为碳负离子机理 [McLennan D J.The carbanion mechanism of olefin-forming elimination. Q Rev Chem Soc, 1967, 21（4）: 490-506.]。该机理包含两个步骤：碳负离子的形成（第一步）和随后的消除（第二步）。（设想 A）步骤 1 是快速可逆的（R 或 rev），步骤 2 是限速步骤（慢速）：（E1cB）$_R$ =（E1cB）$_{rev}$。在这里，反应速率为二级的，限速步骤的速率取决于两种反应物的浓度，即碱（B）和底物（RL）：反应速率 ≈ $k[B][RL]/[BH]$。（设想 B）步骤 1 是慢速且不可逆的（I 或 irr）（限速步骤），步骤 2 是快速的：（E1cB）$_I$ =（E1cB）$_{irr}$。在这里，反应速率是二级，限速步骤速率取决于两种反应物的浓度，即碱（B）和底物（RL）：反应速率 = $k[B][RL]$。（设想 C）步骤 1 是快速的，步骤 2 是限速步骤（慢速）：（E1cB）$_{阴离子}$ =（E1）$_{阴离子}$。在这里，反应速率是一级的，限速步骤速率取决于一种反应物的浓度，也就是底物（RL）：反应速率 ≈ $k[RL]$。

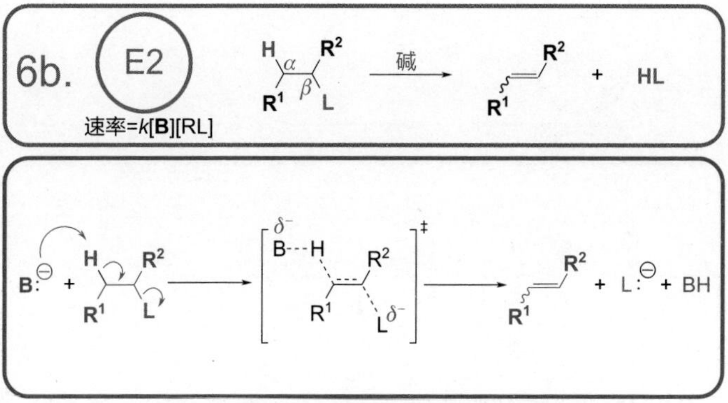

图 6.2 双分子 β- 消除（E2）反应的机理 ❶

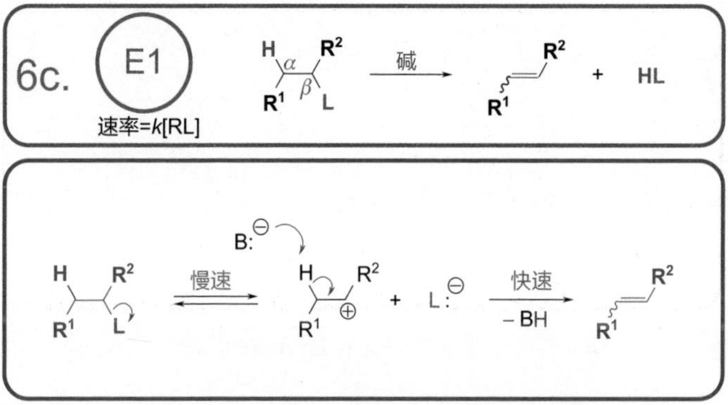

图 6.3 单分子 β- 消除（E1）反应的机理 ❷

---

❶ E2：双分子（2）消除 [elimination **bi**-molecular (**2**)]，即反应速率为二级，限速步骤（即慢速步骤）取决于两种反应物的浓度，在本例中为碱（B）和底物（RL）；反应速率 $=k$ [B][RL]。

❷ E1：单分子（1）消除 [elimination **uni**-molecular（**1**）]，即反应速率为一级，限速步骤（即慢速步骤）取决于一种反应物的浓度，在本例中为底物（RL）；反应速率 $=k$ [RL]。

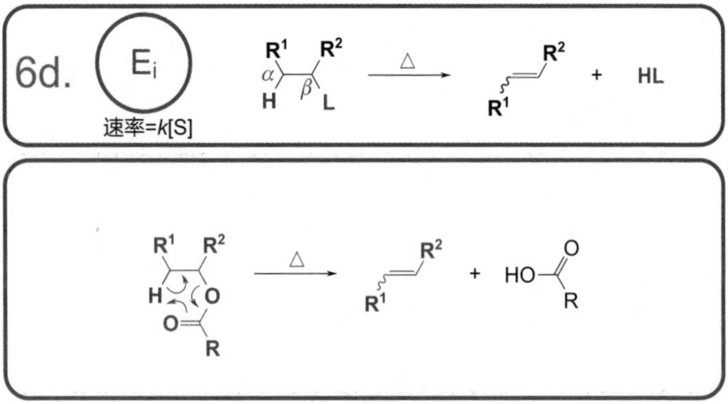

图 6.4　分子内 β- 消除（E_i）反应的机理 ❶

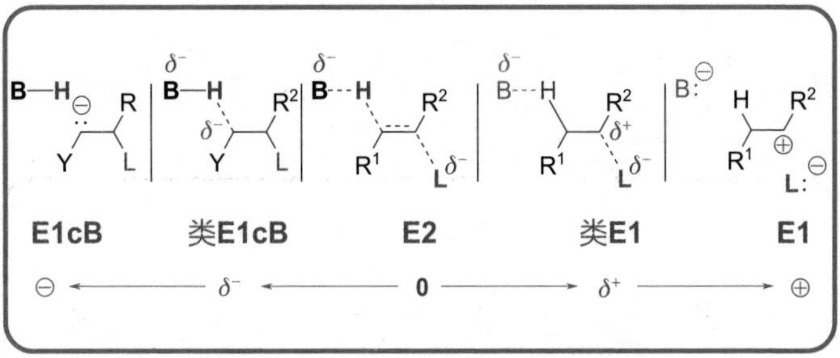

图 6.5　E1cB、E2 和 E1 反应的机理 ❷

---

❶ E_i：分子内消除（elimination internal, intramolecular），反应速率为一级，限速步骤（即慢速步骤）取决于一种反应物的浓度，在本例中为底物（S）；反应速率 $=k\,[\,S\,]$。

❷ E1cB 机理又称为碳负离子机理，其过渡态为带有全负电荷的极大状态。E2 机理是同时发生的，过渡态处在中间。经典的 E2 反应常与 S_N2 反应相竞争，反之亦然。E1 机理与 E1cB 完全相反，其过渡态带正电荷。经典的 E1 反应经常与 S_N1 反应相竞争，反之亦然。

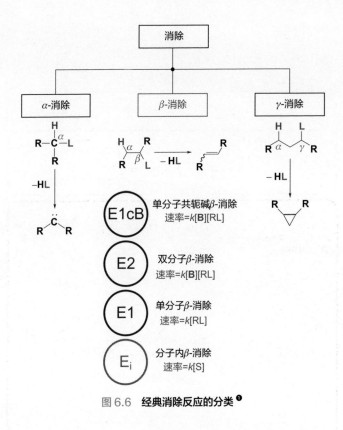

图 6.6 　经典消除反应的分类 ❶

---

❶ 本书中仅介绍主要的 β- 消除实例。

7 酮醇缩合
(acyloin condensation)

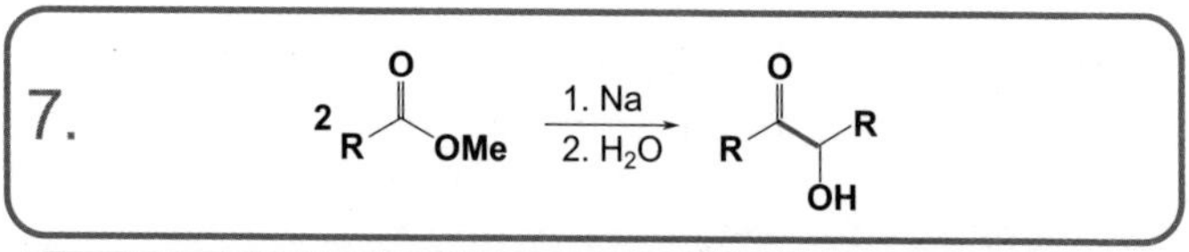

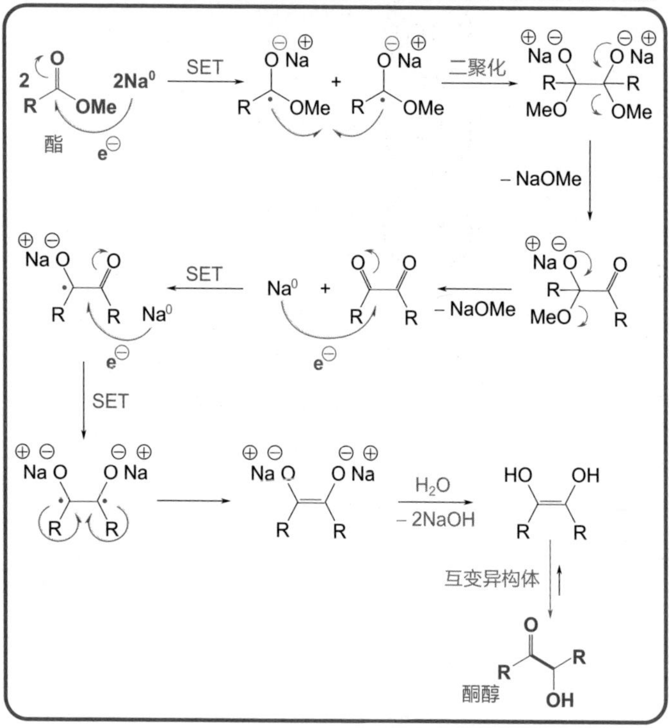

图 7.1 **酮醇缩合的机理**❶

---

❶ 酮醇缩合，又被称为酯的酮醇缩合[译者注：是指两分子酯在金属钠作用下，自身缩合生成 α-羟基酮的反应]，注意：酮醇是指 α-羟基酮。

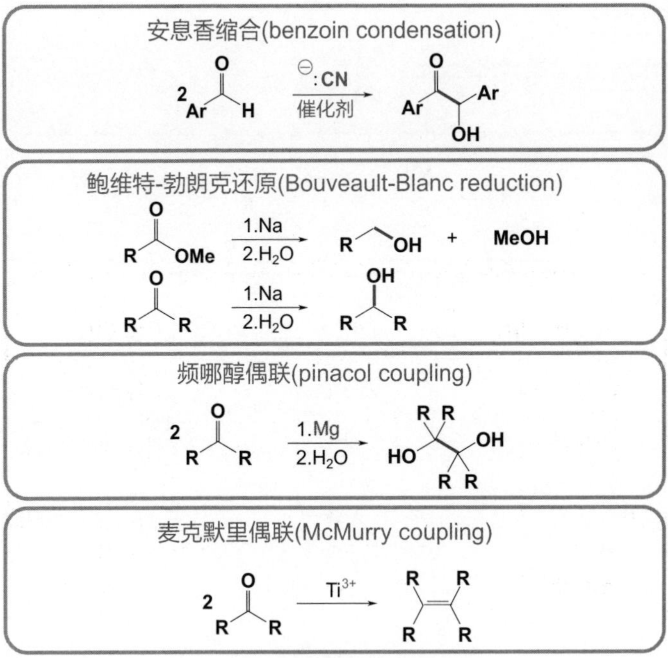
安息香缩合(benzoin condensation)
:CN
催化剂
鲍维特-勃朗克还原(Bouveault-Blanc reduction)
1.Na
2.H2O
MeOH
频哪醇偶联(pinacol coupling)
1.Mg
2.H2O
麦克默里偶联(McMurry coupling)
Ti3+

图 7.2 **酮醇缩合的相关反应** ❶

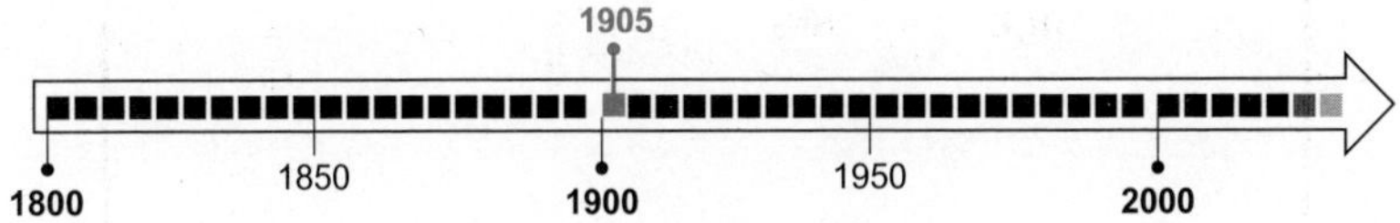
1905
1800
1850
1900
1950
2000

图 7.3 **酮醇缩合的发现** ❷

❶ 与酮醇缩合相关的反应：鲍维特 - 勃朗克还原[1a, 7a]、频哪醇偶联和麦克默里偶联(参见第 57 个反应机理)。安息香缩合(参见第 15 个反应机理)具有不同的机理，但也生成了含有芳香基团的 α- 羟基酮(安息香)。

❷ 该反应大约于 1905 年被首次报道[7b]。

8 炔烃拉链反应
(alkyne zipper reaction)

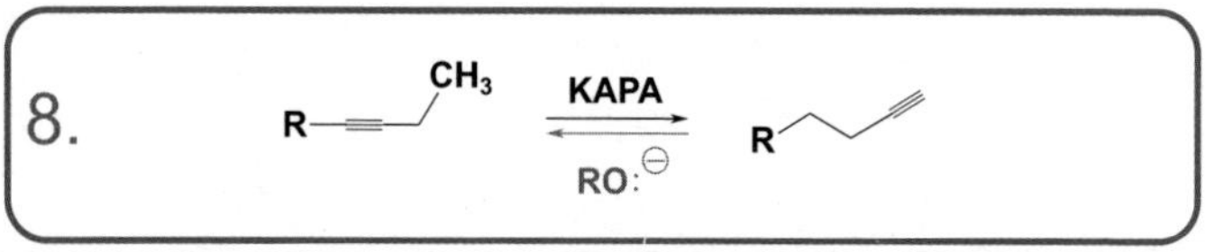

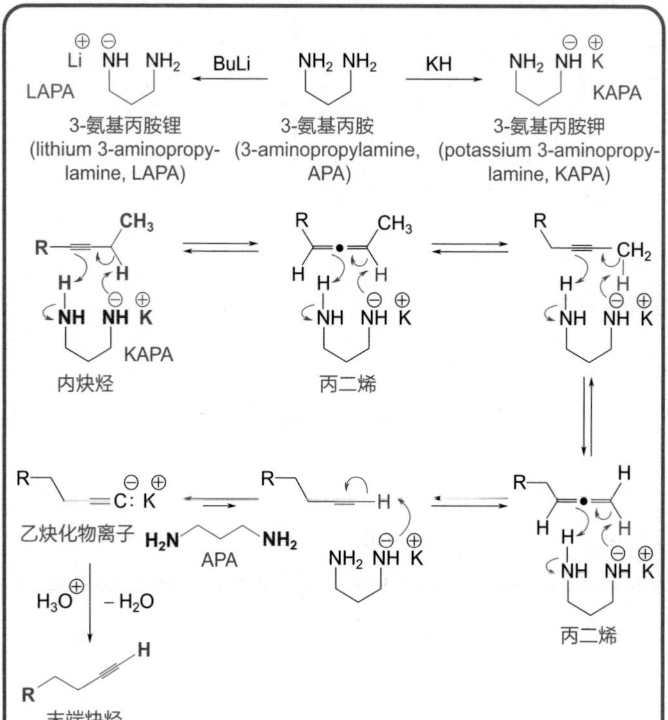

图 8.1 炔烃拉链反应的机理 ❶

---

❶[译者注：炔烃拉链反应是指内炔烃在强碱的作用下经由丙二烯中间体，最终异构化生成末端炔烃的反应。由于异构化的进程就像一条拉链一样，所以该反应俗称为拉链反应。]该反应也称为炔烃异构化反应(alkyne isomerization reaction)或炔烃-丙二烯重排反应(alkyne-allene rearrangement)。

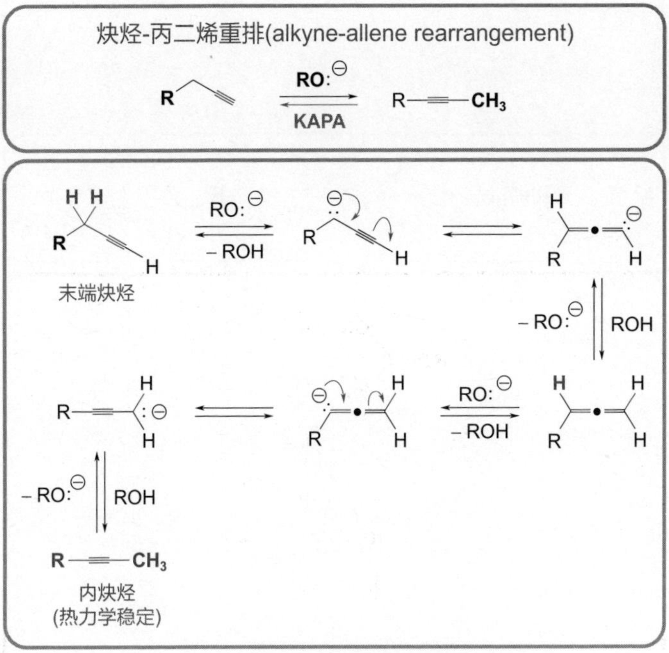

8 炔烃拉链反应

炔烃-丙二烯重排 (alkyne-allene rearrangement)

图 8.2 炔烃-丙二烯重排的机理 ❶

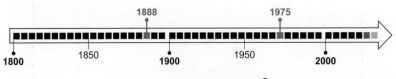

图 8.3 炔烃拉链反应的发现 ❷

---

❶ KAPA 的炔烃拉链反应会生成热力学不稳定的末端炔烃,而经典的炔烃-丙二烯重排反应通常会生成更稳定的内炔烃。两个反应都是可逆的。

❷ 该反应大约于 1888 年被 A. Favorskii[8a, 8b, 8c]首次报道,这里介绍的变体反应大约于 1975 年被首次报道[8d]。

9 阿尔布佐夫反应
(Arbuzov reaction)

9. $(EtO)_3P: + R^1\text{—}X \xrightarrow{-EtX} R^1\text{—}\overset{\overset{O}{\|}}{P}(OEt)_2$ S_N2

亚磷酸酯 + 卤代烷

S_N2, $X = Br, I$

↓

季鏻盐

↓ Δ

↓ $-RX$, S_N2

膦酸酯

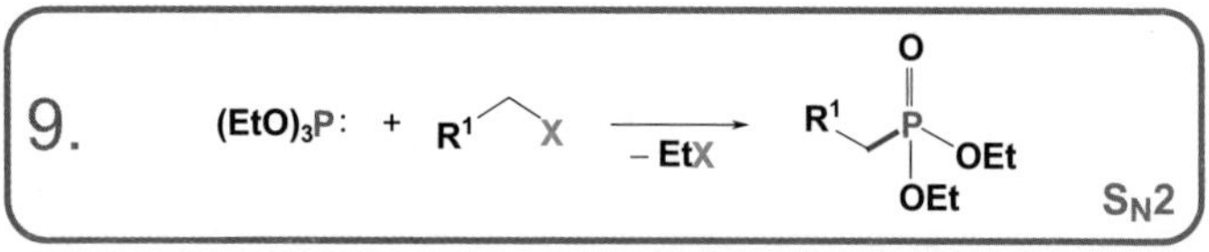

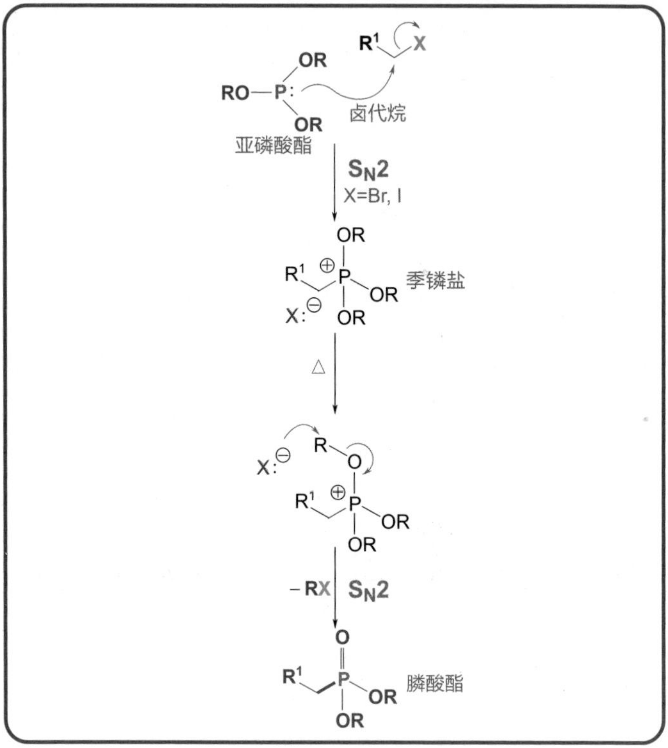

图 9.1 阿尔布佐夫反应的机理 ❶

---

❶[译者注：阿尔布佐夫反应是指三价磷化合物在卤代烷的作用下，转变为五价磷化合物的反应。]阿尔布佐夫反应又被称为米歇尔 - 阿尔布佐夫反应（Michaelis-Aruzov reaction）或米歇尔 - 阿尔布佐夫重排（Michaelis-Aruzov rearranement），属于第 2 个反应机理中介绍的双分子亲核取代（S_N2）的一个实例。

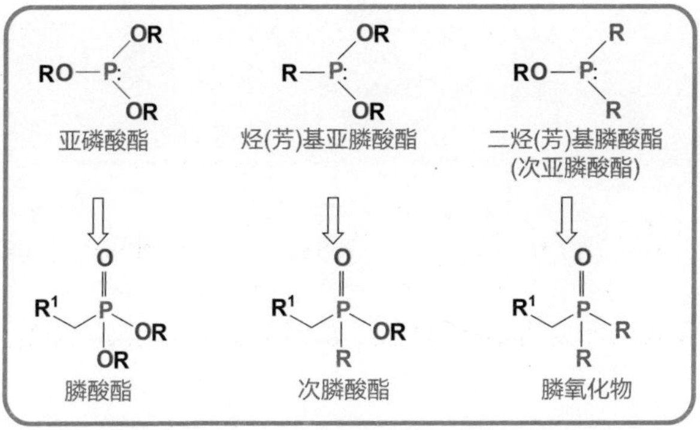

图 9.2 选定的三价（Ⅲ）和五价（Ⅴ）有机磷化合物的命名 ❶

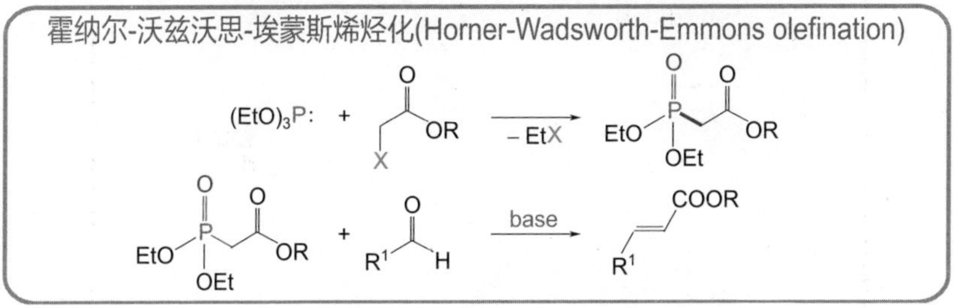

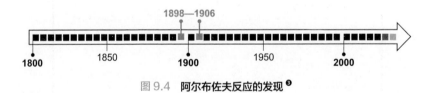

图 9.3 霍纳尔－沃兹沃思－埃蒙斯烯烃化反应 ❷

图 9.4 阿尔布佐夫反应的发现 ❸

---

❶ 译者注："磷"和"膦"的使用经常容易混淆。当磷酸[PO(OH)$_3$]中的一个或两个羟基被取代为烷基或芳基后，所得化合物被称为膦酸。其中，取代一个羟基后被称为膦酸[RPO(OH)$_2$]，取代两个羟基后被称为次膦酸[R$_2$P(O)OH]。同理，当亚磷酸[P(OH)$_3$]中的一个或两个羟基被取代为烷基或芳基后，所得化合物被称为亚膦酸。其中，取代一个羟基后被称为亚膦酸[RP(OH)$_2$]，取代两个羟基后被称为次亚膦酸[R$_2$POH]。复杂有机磷（膦）化物的命名举例：三价（Ⅲ）磷（膦）化合物具有相同的后缀 -ite [如亚磷酸酯 P(OR)$_3$、亚膦酸酯 P(OR)$_2$R、次亚膦酸酯 P(OR)R$_2$]，五价（Ⅴ）磷（膦）化物具有相同的后缀 -ate [如膦酸酯 PO(OR)$_2$R、次膦酸酯 PO(OR)R$_2$] [9a]。

❷ 在阿尔布佐夫反应中产生的膦酸酯是霍纳尔-沃兹沃思-埃蒙斯（Horner-Wadsworth-Emmons，HWE）烯烃化反应（参见第 50 个反应机理）所必需的。

❸ 该反应大约于 1898 年被 Michaelis [9b] 发现，大约于 1906 年被阿尔布佐夫首次报道 [9c]。

10 阿恩特 – 艾斯特尔特合成
(Arndt-Eistert synthesis)

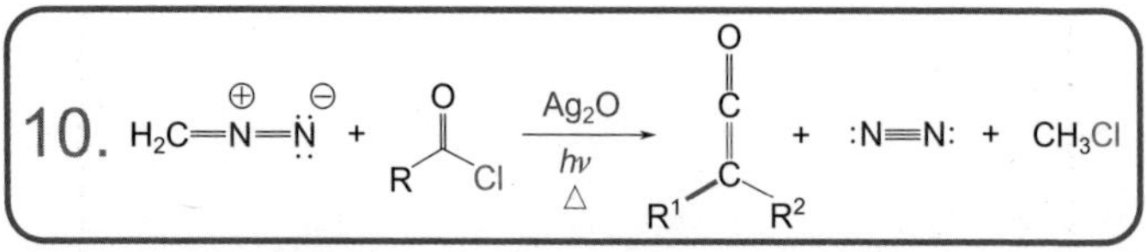

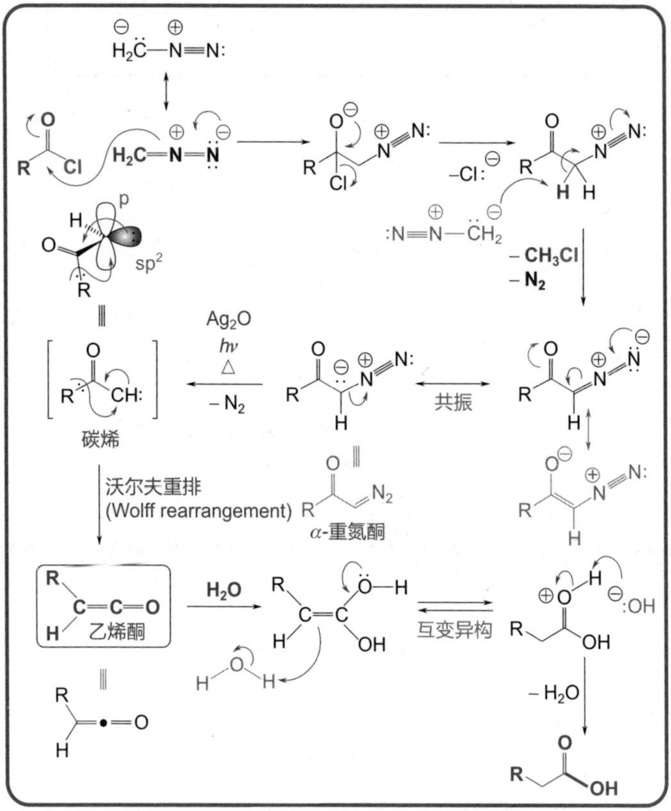

$$10. \quad H_2C=\overset{\oplus}{N}=\overset{\ominus}{N} \; + \; \underset{R}{\overset{O}{\|}}C-Cl \; \xrightarrow[\Delta]{Ag_2O, \; h\nu} \; \underset{R^1 \; R^2}{\overset{O}{\|}}C=C \; + \; :N\equiv N: \; + \; CH_3Cl$$

图 10.1　阿恩特 – 艾斯特尔特合成的机理 ❶

---

❶ 阿恩特 - 艾斯特尔特合成又称为阿恩特 - 艾斯特尔特反应（Arndt-Eistert reaction）(同系化反应)[译者注：是指由羧酸合成多一个碳原子羧酸及其衍生物的反应，羧酸先转变成酰氯，再与重氮甲烷反应生成重氮酮]。沃尔夫重排（Wolff rearrangement）(α- 重氮酮) 是阿恩特 - 艾斯特尔特合成机理的一部分[10a]。

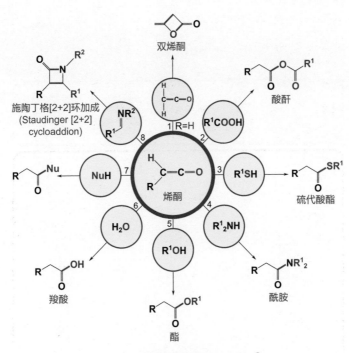

图 10.2　烯酮化合物的合成多样性 ❶

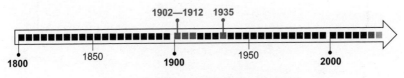

图 10.3　阿恩特 – 艾斯特尔特合成的发现 ❷

---

❶ 在阿恩特 - 艾斯特尔特合成过程中形成的烯酮既可以被各种亲核试剂捕获,也可发生[2+2]环加成反应,包括二聚反应。

❷ 相关反应由沃尔夫(Wolff)于大约 1902—1912 年间首次报道[10a, 10b],以及由阿恩特(Arndt)和艾斯特尔特(Eistert)大约于 1935 年报道[10c]。

11 拜耳 – 维利格氧化
(Baeyer-Villiger oxidation)

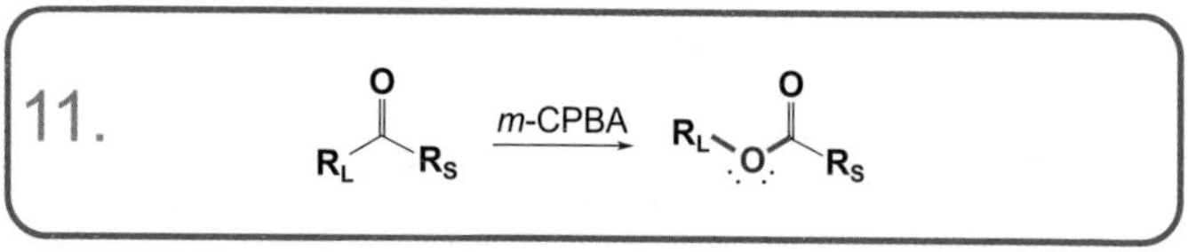

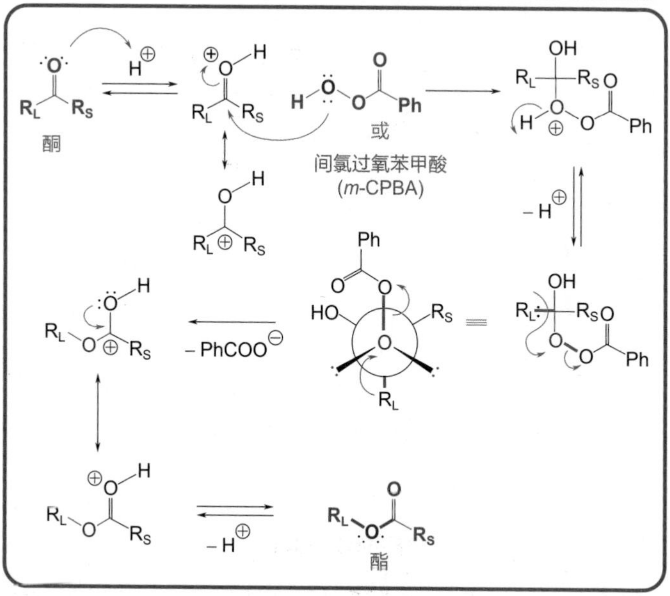

图 11.1　拜耳 – 维利格氧化的机理 ❶

---

❶ 拜耳 - 维利格氧化，也被称为拜耳 - 维利格重排（Baeyer-Villiger rearrangement）[译者注：是指酮在过氧化物（如过氧化氢、过氧化羧酸等）的氧化下，在羰基和一个邻近烃基之间插入一个氧原子，得到相应酯的反应]。

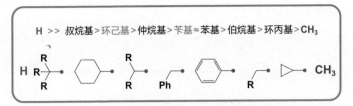

图 11.2　拜耳－维利格氧化过程中基团迁移的顺序 ❶

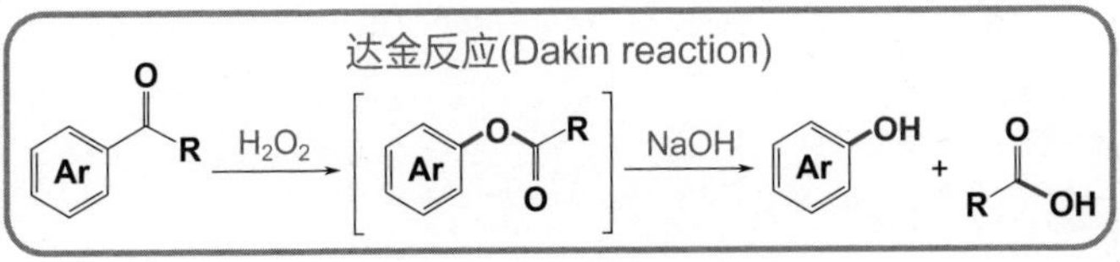

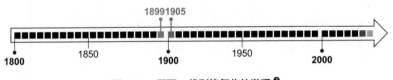

图 11.3　达金反应 ❷

图 11.4　拜耳－维利格氧化的发现 ❸

---

❶ 对于不对称的酮类化合物，基团迁移的顺序至关重要。注意：这种迁移是一种普遍的经验趋势，而不是绝对的规则[1]。

❷ 达金反应[也称为达金氧化（Dakin oxidation）]与拜耳-维利格氧化密切相关，通常生成邻羟基或对羟基苯酚（或具有邻位或对位强 ERG 的苯酚）[11a]。

❸ 该反应大约于 1899 年被首次报道[11b]。鉴于在该领域的成就，约翰·弗里德里希·威廉·阿道夫·冯·拜耳（Johann Friedrich Wilhelm Adolf von Baeyer）于 1905 年获得了诺贝尔化学奖[11c]。

12 巴顿脱羧

(Barton decarboxylation)

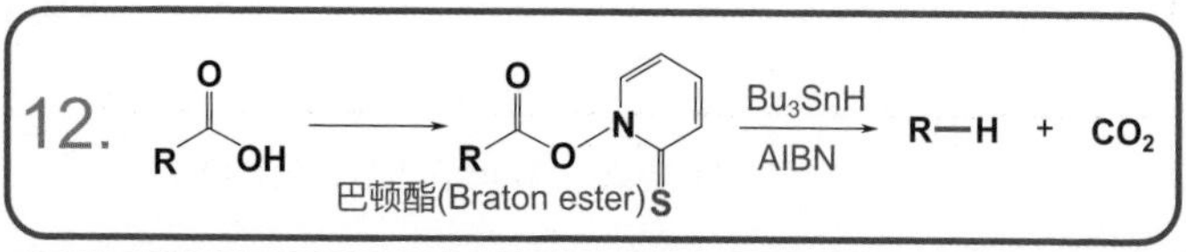
12.
O
R
OH
O
R
O
N
S
巴顿酯(Braton ester)
Bu_3SnH
AIBN
R—H
+
CO_2

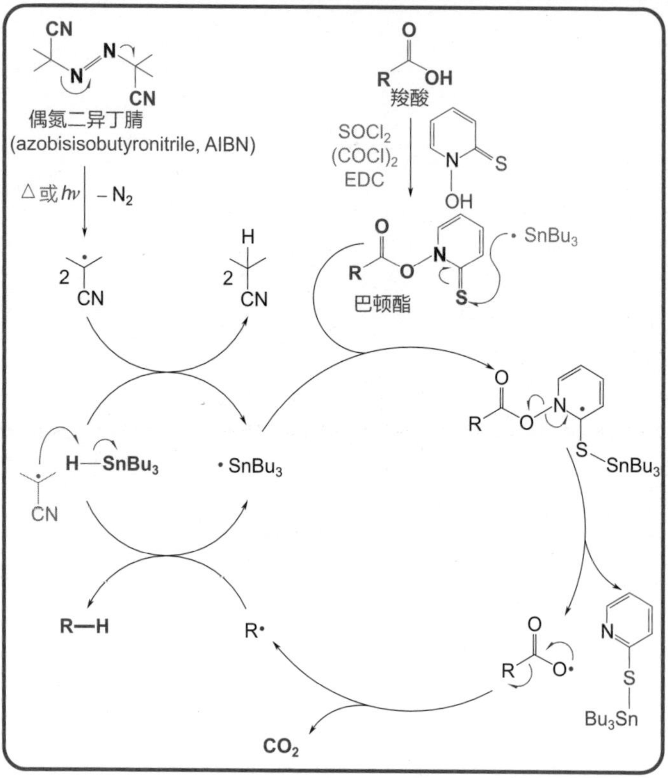
CN
N
N
CN
偶氮二异丁腈
(azobisisobutyronitrile, AIBN)
△或hν
$-N_2$
羧酸
$SOCl_2$
$(COCl)_2$
EDC
OH
$\cdot SnBu_3$
巴顿酯
H—$SnBu_3$
R—H
R·
CO_2
Bu_3Sn

图 12.1 **巴顿脱羧反应的机理**❶

---

❶[译者注：巴顿脱羧是指羧酸转化为巴顿酯，然后进行自由基脱羧生成相应烷烃的反应。]巴顿脱羧是巴顿酯的自由基脱羧反应。

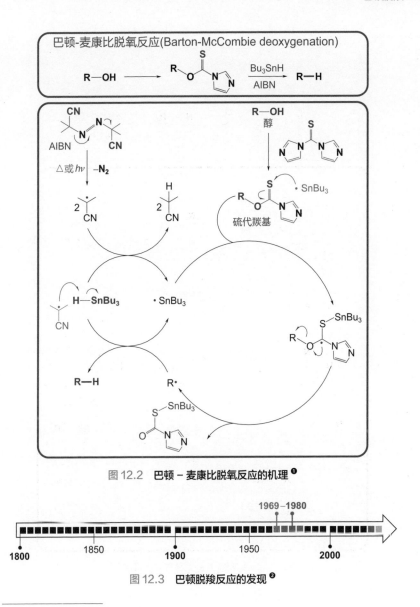

图 12.2 巴顿－麦康比脱氧反应的机理 ❶

图 12.3 巴顿脱羧反应的发现 ❷

---

❶ 巴顿-麦康比脱氧反应是硫代羰基的自由基脱氧反应：O,O-硫代碳酸盐 ROC（S）OR；S,O-二硫代碳酸酯 = 黄原酸盐 ROC（S）SR；O-硫代氨基甲酸酯 ROC（S）NR_2。

❷ 脱羧反应大约于 1980—1985 年被首次报道[12a, 12b]，脱氧反应大约于 1975—1980 年被首次报道[12c, 12d]。鉴于在该领域的成就，Derek H. R. Barton 与 Odd Hassel 于 1969 年共同获得了诺贝尔化学奖[12e]。

13 贝里斯 – 希尔曼反应
(Baylis–Hillman reaction)

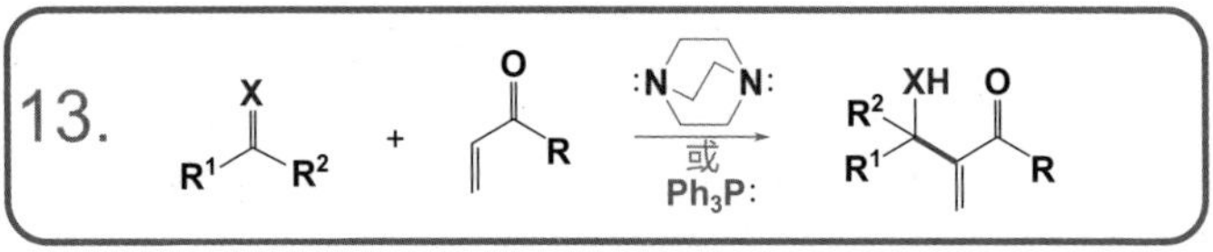

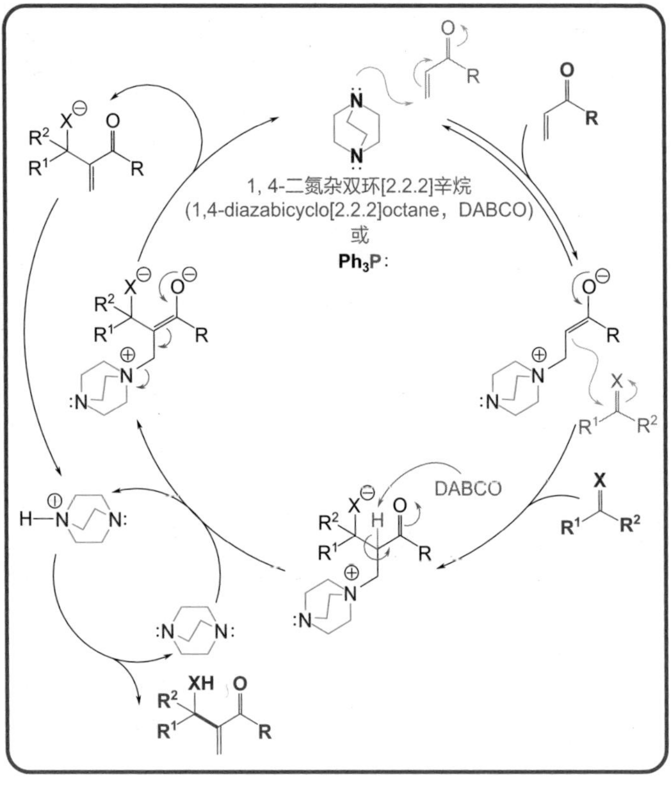

图 13.1　贝里斯 – 希尔曼反应的机理[1]

---

[1] 贝里斯 - 希尔曼反应，又称为森田 - 贝里斯 – 希尔曼反应（Morita-Baylis–Hillman reaction）[译者注：是指 α, β- 不饱和化合物与亲电试剂（醛、酮）在催化剂作用下生成烯烃 α- 位加成产物的反应]。

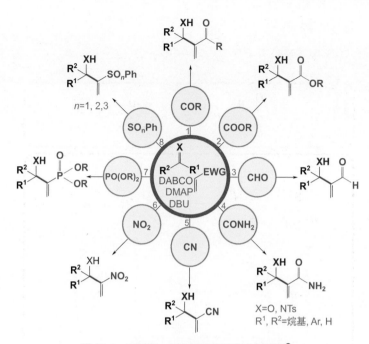

图 13.2　贝里斯-希尔曼反应的合成多样性 ❶

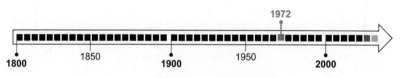

图 13.3　贝里斯-希尔曼反应的发现 ❷

---

❶ 贝里斯-希尔曼反应存在多种变体反应，具体取决于 EWG（迈克尔受体）和羰基化合物（亲电试剂）的性质。注意：当 X=NR 时，该反应称为氮杂-贝里斯-希尔曼反应（aza-Baylis-Hillman reaction）。

❷ 该反应大约于 1972 年被首次报道[13]。

14 贝克曼重排
(Beckmann rearrangement)

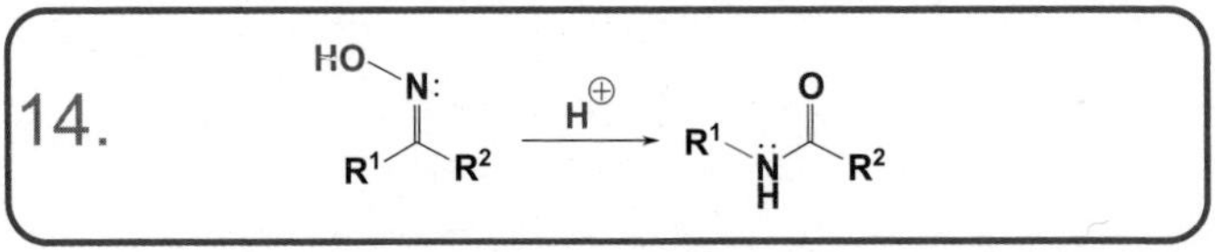

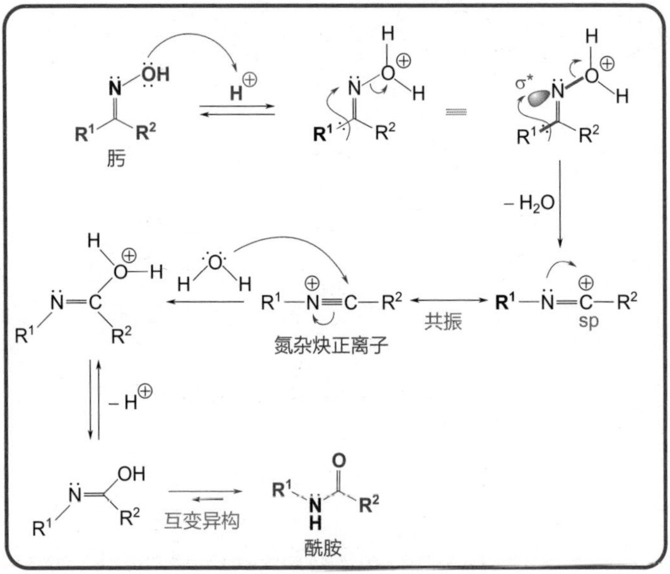

图 14.1 贝克曼重排的机理❶

---

❶[译者注：贝克曼重排是指醛肟或酮肟在酸催化下生成 N- 取代酰胺的亲核重排反应。] 贝克曼重排很少被称为贝克曼肟 - 酰胺重排（Beckmann oxime-amide rearrangement）。

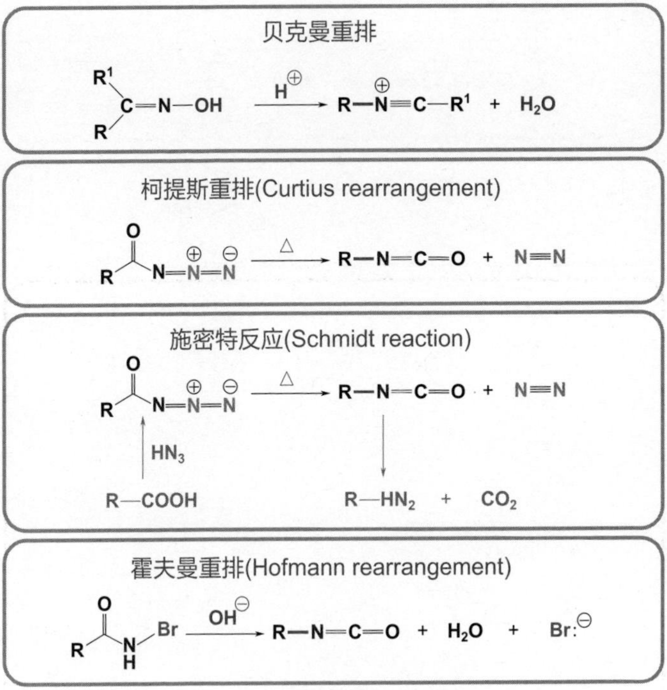

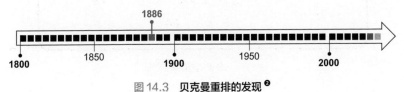

图 14.2　贝克曼重排的相关反应 ❶

图 14.3　贝克曼重排的发现 ❷

---

❶ 多个反应与贝克曼重排机理相关：柯提斯重排、施密特反应、霍夫曼重排以及洛森重排（Lossen rearrangement）（参见第 31 个反应机理）。第一个贝克曼重排的实例强调肟重排为氮离子；在其他实例中，关键步骤是将氮烯（由羰基衍生物形成）重排为异氰酸酯。

❷ 该反应大约于 1886 年被首次报道[14]。

15 安息香缩合
(benzoin condensation)

$$2\ \text{Ar-CHO} \xrightarrow[\text{催化剂}]{^{\ominus}:\text{CN}} \text{Ar-CO-CH(OH)-Ar}$$

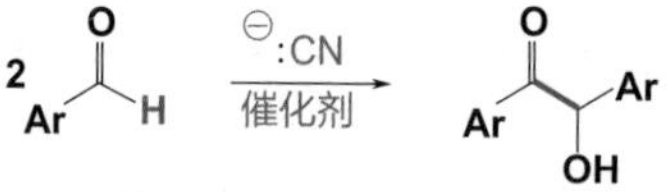

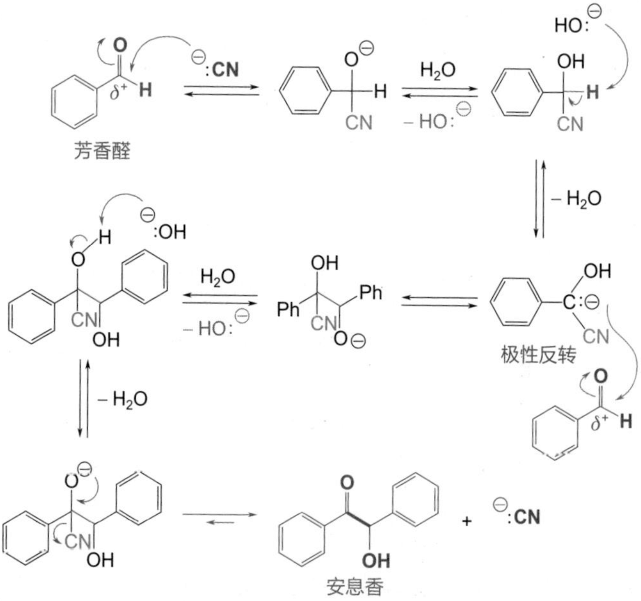

图 15.1 安息香缩合的机理 ❶

---

❶ [译者注：安息香缩合是指芳香醛在氰基源亲核催化剂的催化下，自身发生缩合反应生成安息香产物的反应。]安息香缩合是有机化学中最古老的反应之一。

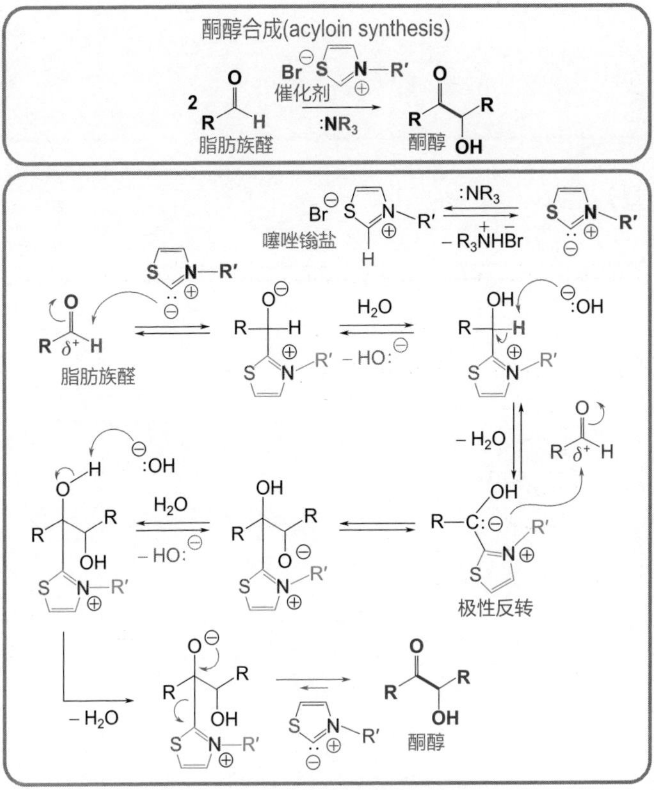

15 安息香缩合

[图：酮醇合成(acyloin synthesis) 与采用噻唑鎓盐的酮醇合成机理示意]

图 15.2 采用噻唑鎓盐的酮醇合成的机理 ❶

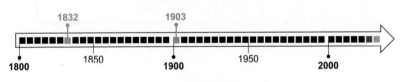

图 15.3 安息香缩合的发现 ❷

---

❶ 安息香缩合反应涉及两分子芳香醛，在氰化物离子（cyanide ion）催化下生成芳香 α-羟基酮（安息香）。酮醇的合成是在噻唑鎓盐催化[15a, 15b]下，由两个脂肪醛缩合生成脂肪族（或混淆）α-羟基酮（酮醇）。不应把酮醇合成与酮醇缩合相混淆（参见第 7 个反应机理）。

❷ 该反应大约于 1832 年被首次报道，其机理大约于 1903 年被提出[15c, 15d]。

16 苯炔机理
(benzyne mechanism)

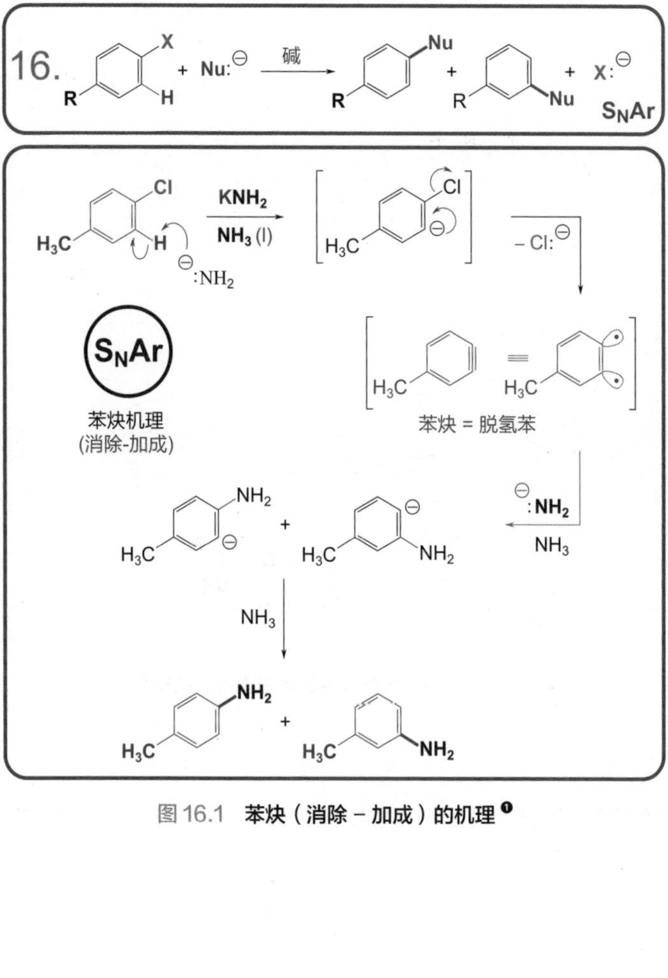

图 16.1 苯炔（消除－加成）的机理 ❶

---

❶ 苯炔机理是最基本的芳香亲核取代机理之一，也被称为消除-加成机理或加成-消除机理（参见第 4 个反应机理），与 S_NAr（S_N2Ar）相反。

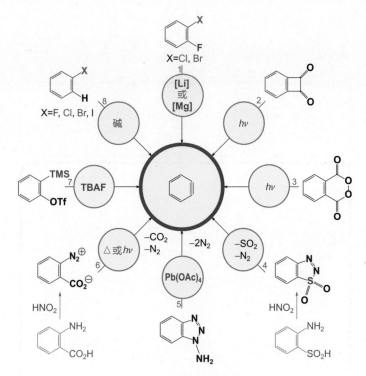

图 16.2　合成苯炔的多种方法 ❶

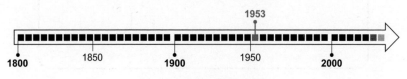

图 16.3　苯炔机理的发现 ❷

---

❶ 自该机理被首次发现以来,已开发了多种方法用于苯炔中间体(芳炔)的合成。注意:苯炔(芳炔)也被称为脱氢苯(脱氢芳烃)[16a, 16b]。

❷ 该机理大约于 1953 年被首次报道[16c]。

17 伯格曼环化
(Bergman cyclization)

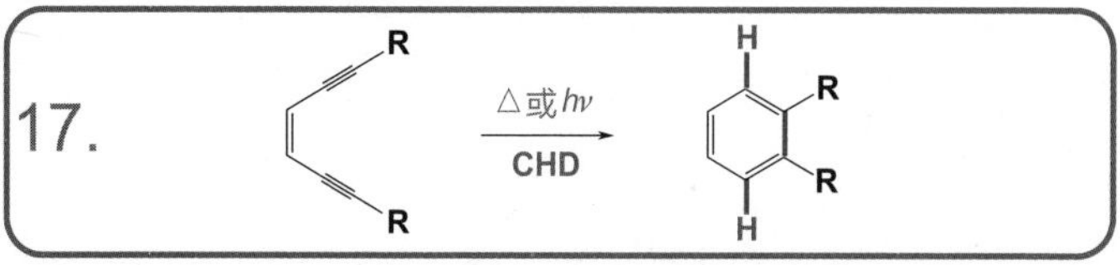

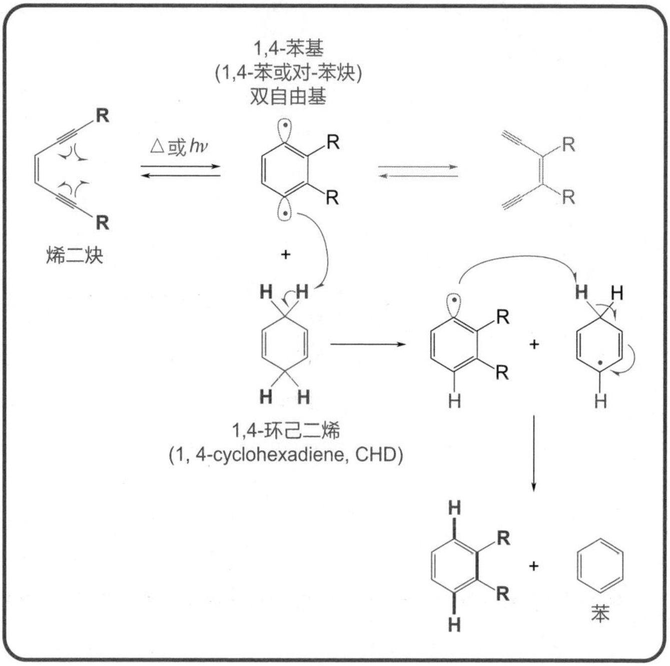

图 17.1 **伯格曼环化的机理** ❶

---

❶伯格曼环化，又称伯格曼反应（Bergman reaction）[译者注：是指共轭的烯二炔通过分子内环化生成1,4-苯双自由基（对苯炔）或其类似物的一类环化反应]。

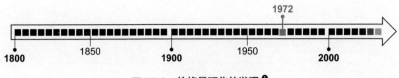

图 17.2　**伯格曼环化的发现** ❶

---

❶ 该反应大约于 1972 年被首次报道[17]。

18 伯奇还原
(Birch reduction)

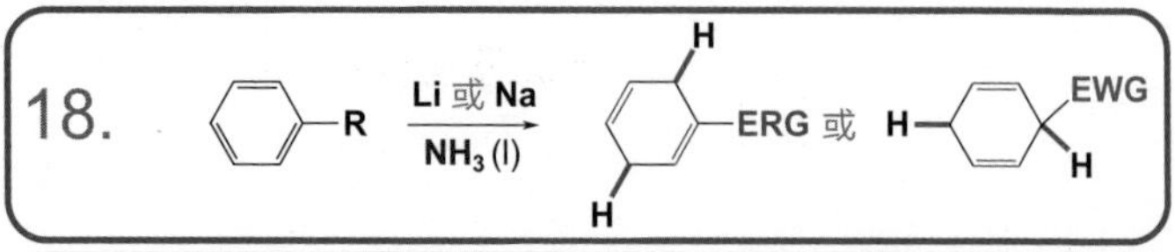

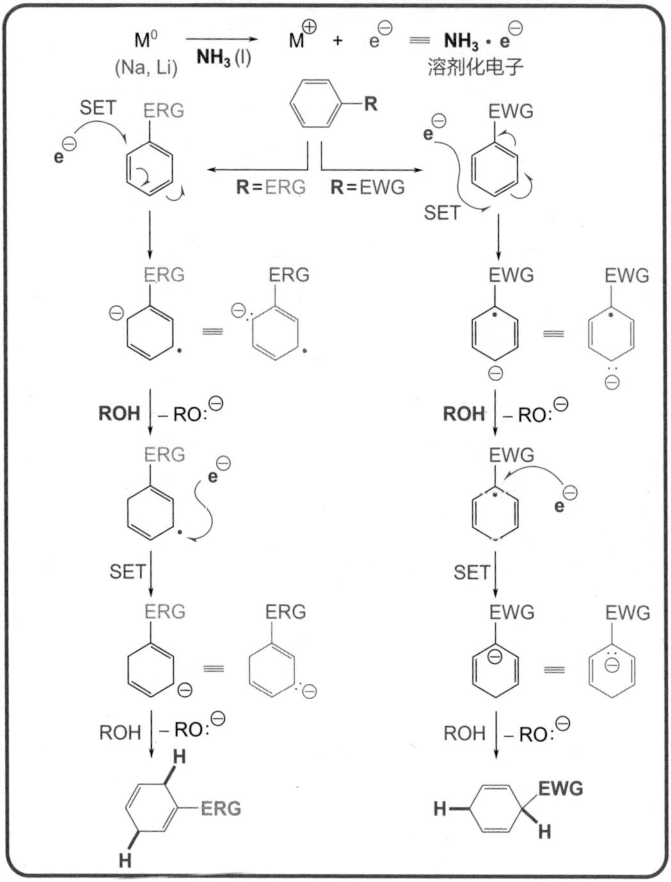

图 18.1 **伯奇还原的机理** ❶

---

❶ [译者注：伯奇还原是指芳香环通过碱金属（Li、Na、K）液氨溶液在醇存在下进行 1,4- 位还原得到非共轭的环己二烯或其他不饱和杂环的反应。] 伯奇还原机理的第一步为单电子转移（SET）（参见第 5 个反应机理）。所形成产物的区域选择性取决于取代基的性质（ERG 和 EWG）。

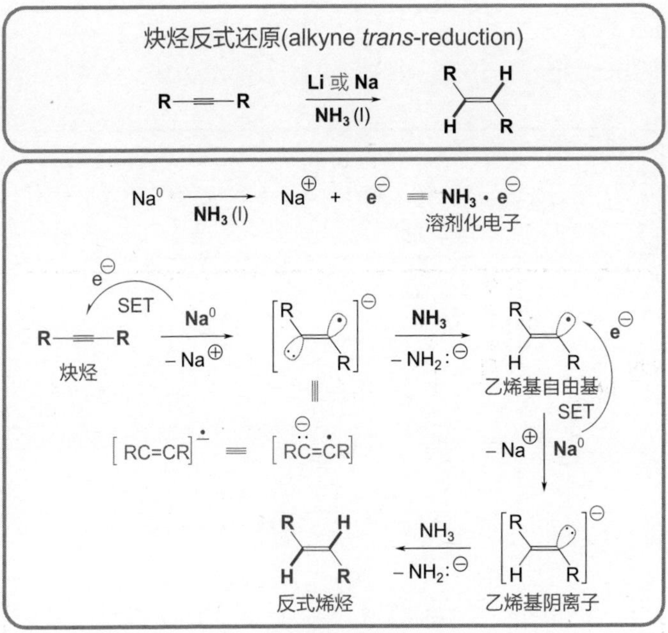

炔烃反式还原 (alkyne *trans*-reduction)

$$R-\!\!\!\equiv\!\!\!-R \xrightarrow[NH_3(l)]{Li \text{ 或 } Na} \begin{array}{c} R \\ H \end{array}\!\!=\!\!\begin{array}{c} H \\ R \end{array}$$

图 18.2 炔烃反式还原的机理 ❶

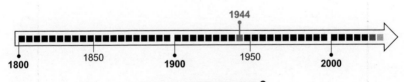

图 18.3 伯奇还原的发现 ❷

---

❶ 炔烃反式还原（alkyne *trans*-reduction）（炔烃金属还原）的机理与伯奇还原很相似。注意：在伯奇还原条件下，炔烃被还原为反式烯烃（*trans*-alkene）[18a, 18b]。

❷ 该反应大约于 1944 年被首次报道[18c]。

19 比施勒 – 纳皮耶拉尔斯基环化
(Bischler-Napieralski cyclization)

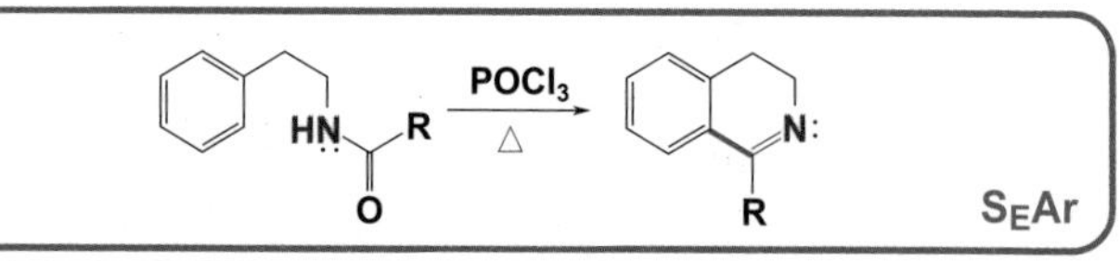

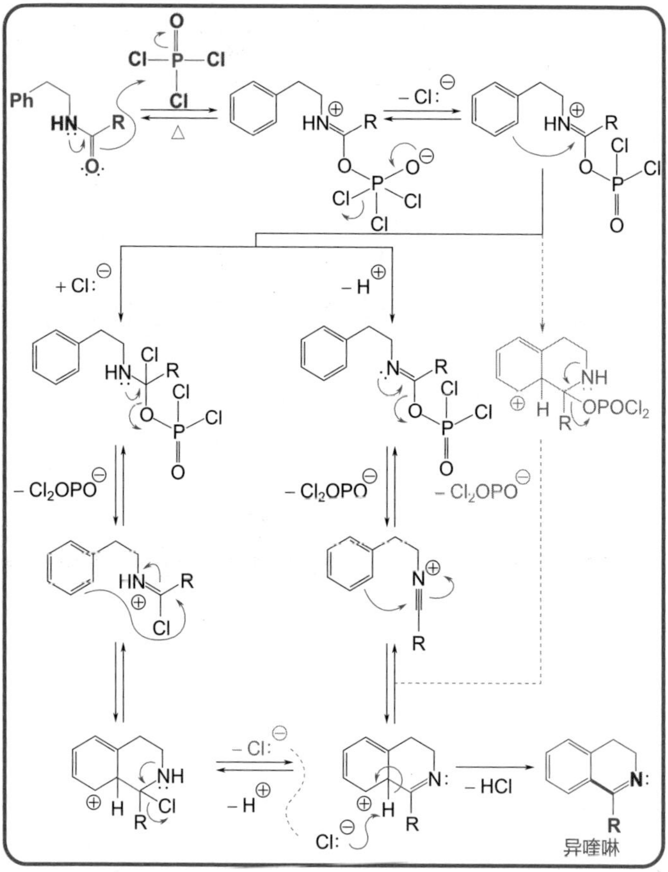

图 19.1 比施勒 – 纳皮耶拉尔斯基环化反应的机理 ❶

---

❶ 比施勒 - 纳皮耶拉尔斯基环化，又称为比施勒 - 纳皮耶拉尔斯基反应（Bischler-Napieralski reaction）[译者注：是指 β- 芳基乙基酰胺在氯代试剂的作用下，发生分子内亲电取代环合生成二氢异喹啉类化合物的反应]。该反应是芳香亲电取代的一个经典实例（参见第 3 个反应机理，芳香正离子机理或 S_EAr）。

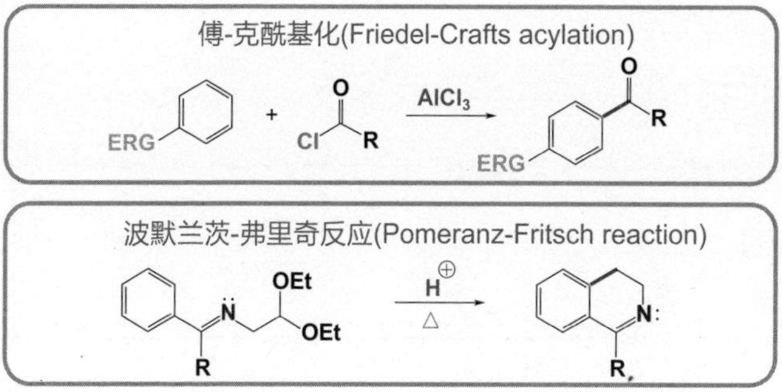

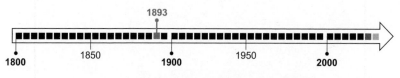

图 19.2　与比施勒-纳皮耶拉尔斯基环化相关的反应 ❶

图 19.3　比施勒-纳皮耶拉尔斯基环化反应的发现 ❷

---

❶ 多个人名反应与比施勒-纳皮耶拉尔斯基环化有关：傅-克酰基化和烷基化反应（参见第 39 个反应机理），以及密切相关的波默兰茨-弗里奇反应，这些都是合成异喹啉的其他方法[19a, 19b]。

❷ 该反应大约于 1893 年被首次报道[19c]。

20 布朗硼氢化
(Brown hydroboration)

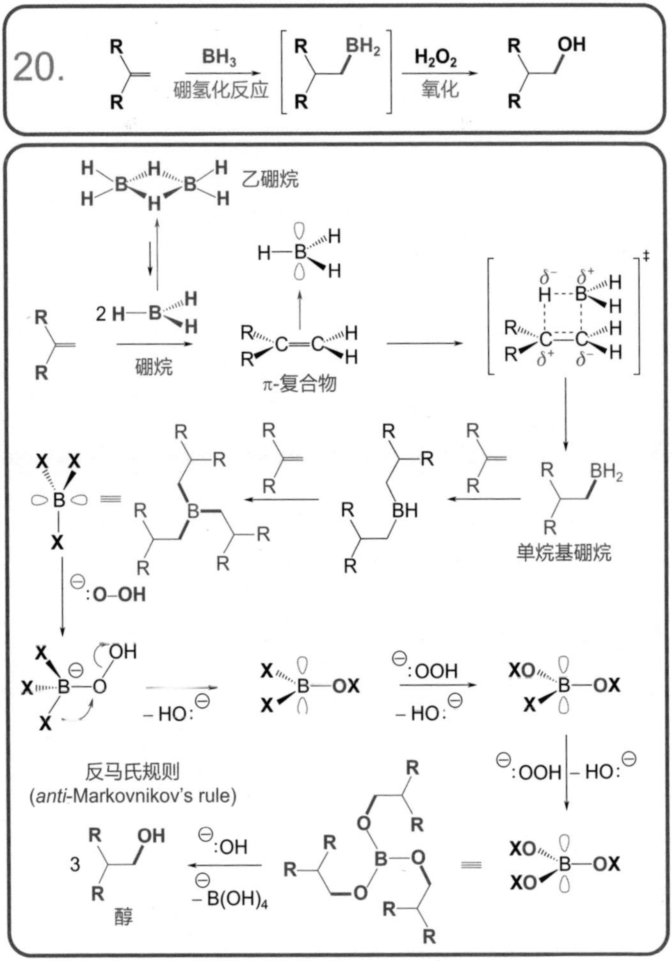

图 20.1 布朗硼氢化的机理 ❶

---

❶ 布朗硼氢化，又称为硼氢化氧化（hydroboration-oxidation）[译者注：是指乙硼烷在醚类溶液中解离成的甲硼烷的 B—H 键与烯烃、炔烃不饱和键加成，生成有机硼化合物的反应]。该机理被认为是协同机理，通常生成反马氏规则产物（参见第 52 个反应机理）。

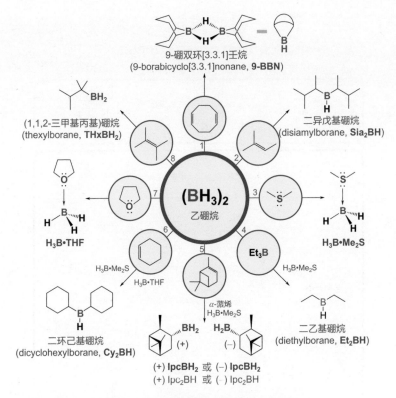

图 20.2　由乙硼烷形成的各种硼烷衍生物 ❶

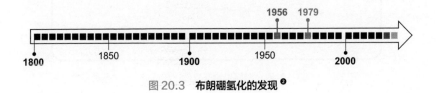

图 20.3　布朗硼氢化的发现 ❷

---

❶ 存在许多硼烷复合物（$BH_3·X$）的实例：单烷基硼烷（RBH_2）；二烷基硼烷（R_2BH）试剂，由二硼烷（B_2H_6）经硼氢化反应制得；9-硼双环[3.3.1]壬烷（9-BBN）试剂（其中最重要的一种）[20a]。

❷ 该反应大约于 1956 年被首次报道[20b]。1979 年，Herbert C. Brown 与 Georg Wittig 因在硼化学领域的成就而共同获得了诺贝尔化学奖[20c]。

21 布赫瓦尔德 – 哈特维希交叉偶联
(Buchwald-Hartwig cross coupling)

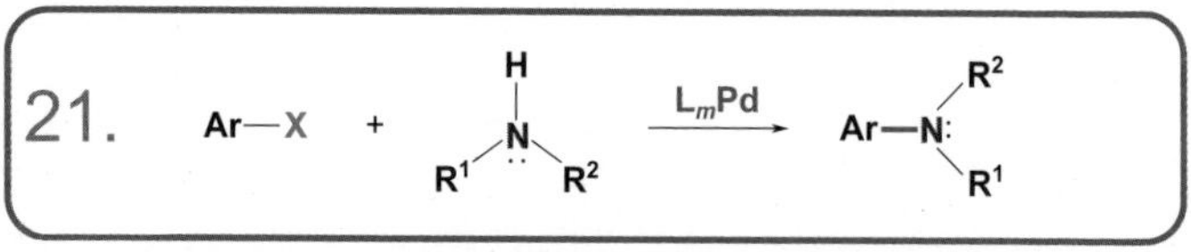

21. $Ar-X$ + $HN(R^1)(R^2)$ $\xrightarrow{L_mPd}$ $Ar-N(R^2)(R^1)$

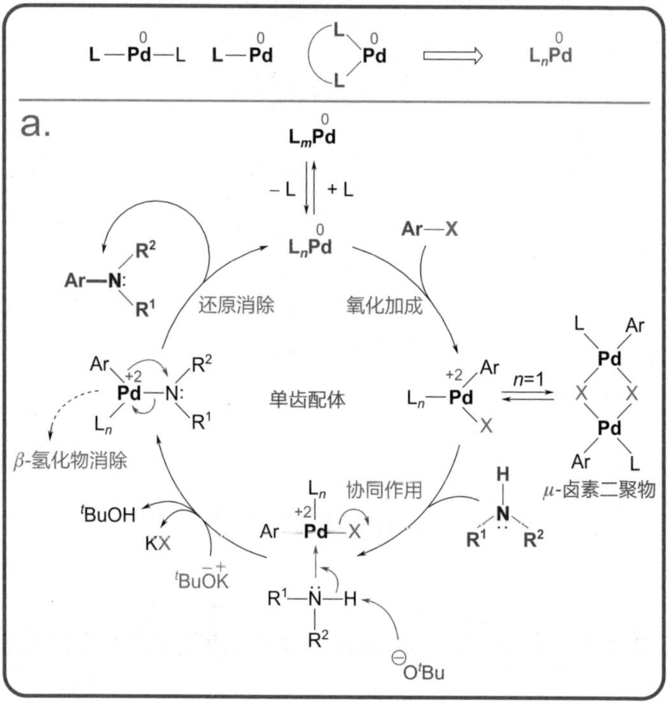

a.

$L-Pd-L$ $L-Pd$ $L\overset{L}{\underset{L}{Pd}}$ $\Longrightarrow$ L_nPd^0

L_mPd^0

$-L \rightleftharpoons +L$

L_nPd^0

$Ar-X$ 氧化加成

$L_n\overset{+2}{Pd}\begin{smallmatrix}Ar\\X\end{smallmatrix}$ $\xrightleftharpoons{n=1}$ $\begin{smallmatrix}L\\Pd\\Ar\end{smallmatrix}\overset{Ar}{\underset{X}{\diamondsuit}}\begin{smallmatrix}Pd\\L\end{smallmatrix}$ μ-卤素二聚物

单齿配体

$Ar-\overset{+2}{\underset{L_n}{Pd}}-N\begin{smallmatrix}R^2\\R^1\end{smallmatrix}$ β-氢化物消除

还原消除 → $Ar-N(R^1)(R^2)$

^tBuOH KX $^tBuO^-K^+$

$Ar-\overset{+2}{\underset{L_n}{Pd}}-X$ 协同作用 $HN(R^1)(R^2)$

$R^1-\overset{+}{N}-H$
     R^2

$^-O^tBu$

图 21.1 **布赫瓦尔德 – 哈特维希交叉偶联的机理（单齿配体）**❶

---

❶[译者注：布赫瓦尔德 – 哈特维希交叉偶联是指金属钯催化的卤代芳烃（常被三氟甲磺酸的酚酯代替）与胺（伯胺或仲胺）或醇之间的偶联反应。]布赫瓦尔德 – 哈特维希交叉偶联（胺化）是一种钯催化的交叉偶联反应，使卤代芳烃与胺形成 C—N 键。其机理各不相同，但通常取决于底物和配体。图中仅显示了一个在单齿配体存在下可能发生的简单实例。

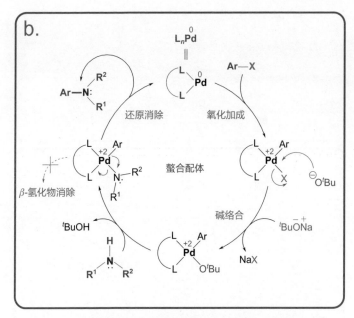

图 21.2 布赫瓦尔德 – 哈特维希交叉偶联的机理（螯合配体）❶

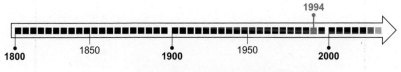

图 21.3 布赫瓦尔德 – 哈特维希交叉偶联的发现 ❷

---

❶ 图中仅显示了一个在螯合配体存在下可能发生的简单实例。
❷ 该反应大约于 1994 年被首次报道[21]。

22 坎尼扎罗反应
(Cannizzaro reaction)

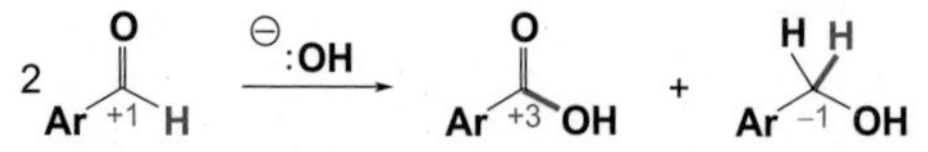

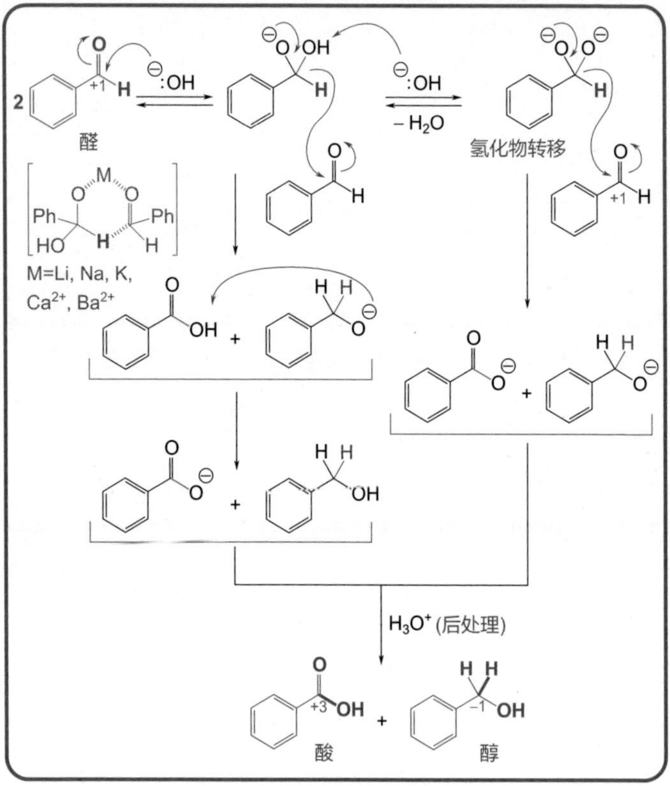

图 22.1 **坎尼扎罗反应的机理** ❶

---

❶[译者注：坎尼扎罗反应是指无 α-活泼氢的醛在强碱作用下发生分子间氧化还原反应，生成一分子羧酸和一分子醇的歧化反应。]坎尼扎罗反应是有机化学中最古老的反应之一，很少被称为坎尼扎罗歧化（Cannizzaro disproportionation）反应。

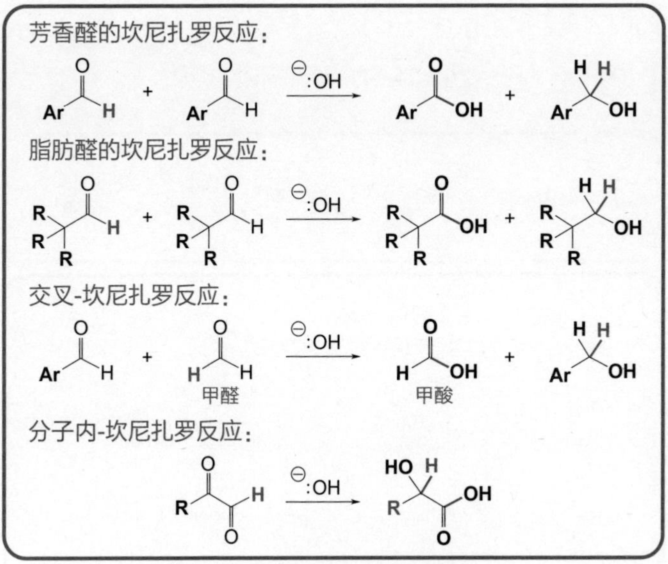

图 22.2　坎尼扎罗反应的变体反应 ❶

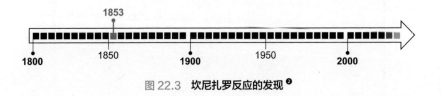

图 22.3　坎尼扎罗反应的发现 ❷

---

❶ 坎尼扎罗反应具有多种变体反应：与不含 α-氢原子的芳香醛和脂肪醛的坎尼扎罗反应、交叉-坎尼扎罗反应，以及分子内坎尼扎罗反应[1]。

❷ 该反应大约于 1853 年被首次报道[22]。

23 陈－埃文斯－兰交叉偶联
(Chan-Evans-Lam cross coupling)

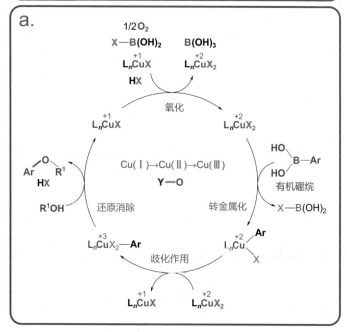

图 23.1　陈－埃文斯－兰交叉偶联的机理（Y=O）❶

---

❶ 陈-埃文斯-兰交叉偶联，又称为陈-兰交叉偶联（Chan-Lam cross coupling），是一种铜催化的芳基硼酸和醇或胺形成 C—O 和 C—N 键的交叉偶联反应。其机制尚不明确，通常非常依赖于底物和配体。图中仅显示了一个醚化反应（C—O 键形成，Y=O）可能机理的简单实例[23a, 23b]。

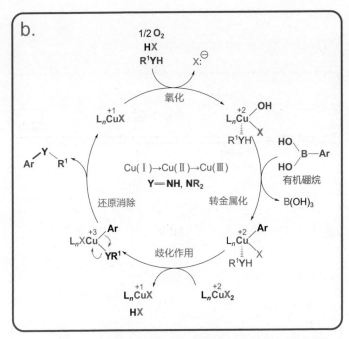

图 23.2 陈-埃文斯-兰交叉偶联的机理（Y=NH、NR$_2$）❶

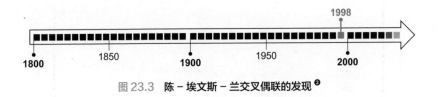

图 23.3 陈-埃文斯-兰交叉偶联的发现 ❷

---

❶ 该反应的机理尚不完全明确，通常高度取决于底物和配体。为了便于理解，图中仅显示了一个在胺化反应中的简单实例（C—N 键形成，Y=NH、NR$_2$）[23c]。

❷ 该反应大约于 1998 年被首次报道[23d, 23e, 23f]。

24 齐齐巴宾氨基化
(Chichibabin amination)

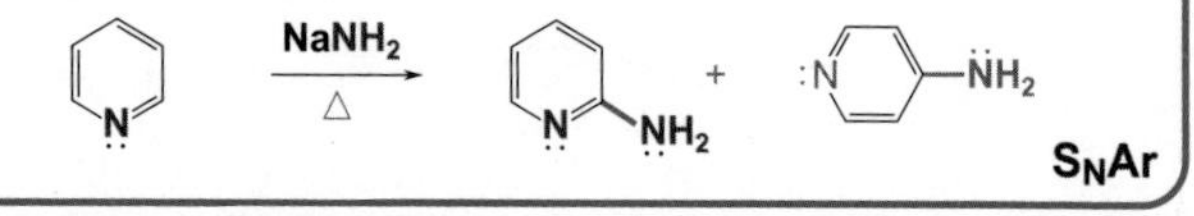

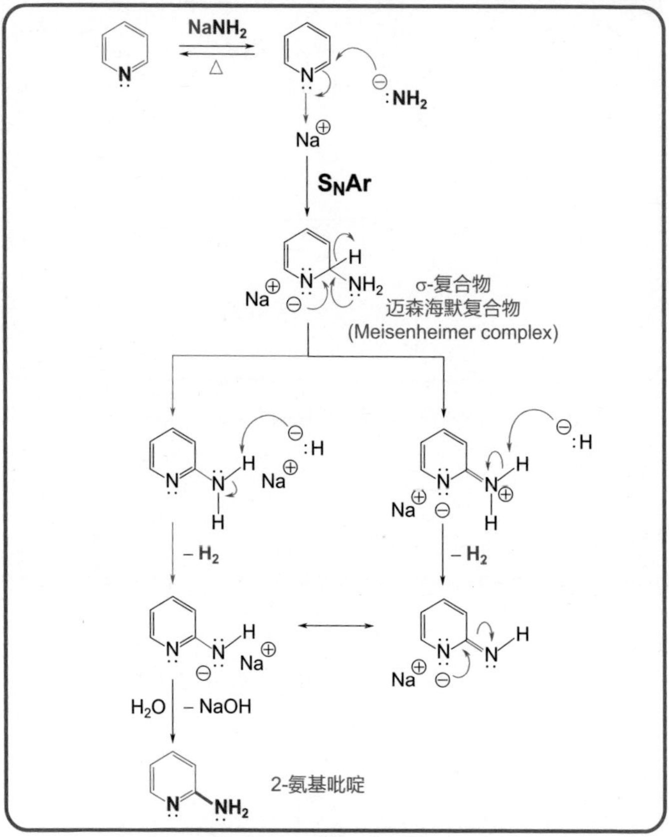

图 24.1 **齐齐巴宾氨基化的机理** ❶

---

❶ 齐齐巴宾氨基化，又称齐齐巴宾反应（Chichibabin reaction）[译者注：是指吡啶或其衍生物等含氮杂环碱类化合物与碱金属的氨基化合物在加热时发生胺化生成相应的氨基衍生物的反应]。齐齐巴宾氨基化是一个典型的芳香亲核取代反应，具体而言，经历了加成 - 消除机理：S_NAr（S_N2Ar）（参见第 4 个反应机理）。

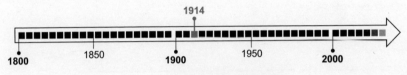

图 24.2　**齐齐巴宾氨基化的发现** ❶

---

❶ 该反应大约于 1914 年被首次报道[24]。

25 克莱森缩合
(Claisen condensation)

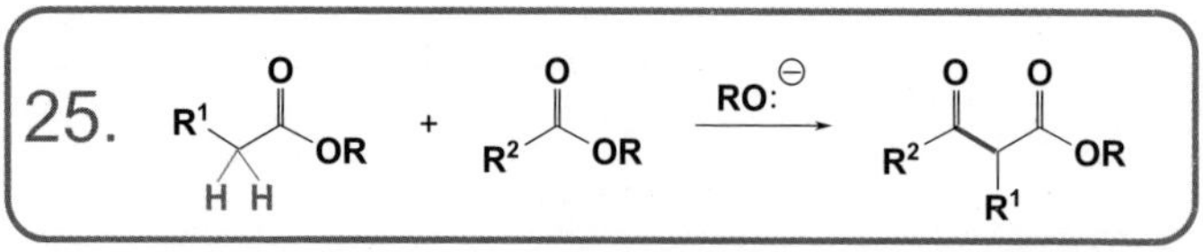

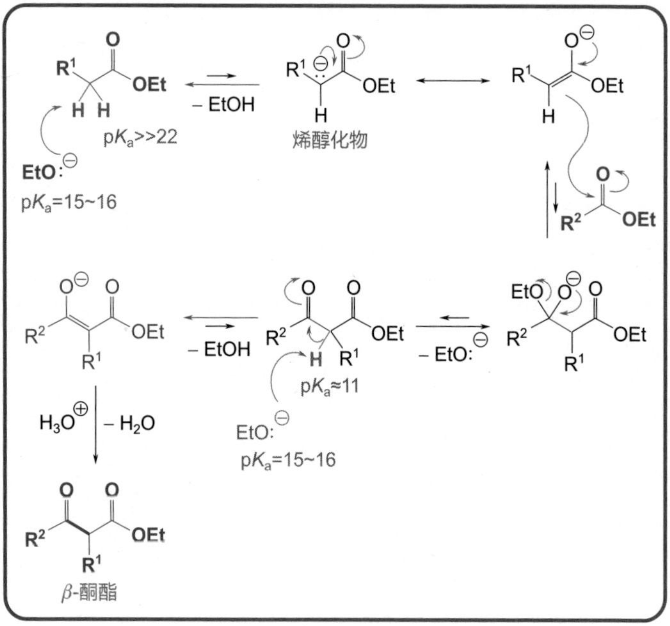

图 25.1 **克莱森缩合的机理** ❶

---

❶[译者注：克莱森缩合是指含有 α-H 的酯在醇钠等碱作用下发生缩合，失去一分子醇生成 β-酮酸酯的反应。]克莱森缩合是一种酯与另一羰基化合物之间的缩合反应，该化合物含有两个可烯化的氢原子（α-H）。

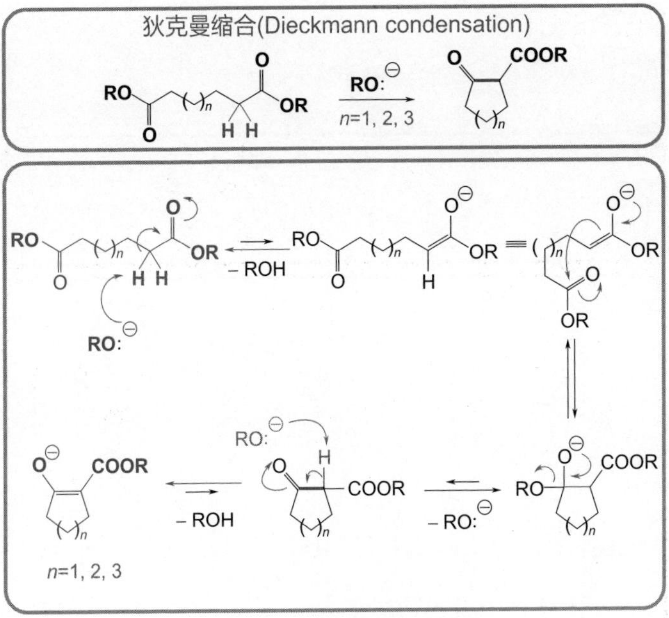

图 25.2 狄克曼缩合的机理 ❶

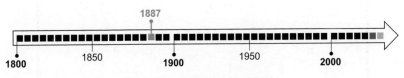

图 25.3 克莱森缩合的发现 ❷

---

❶ 狄克曼缩合是分子内的克莱森缩合，二者机理几乎相同。狄克曼缩合是形成五元环、六元环和七元环的理想方法[25a]。

❷ 该反应大约于 1887 年被首次报道[25b]。

26 克莱森重排
(Claisen rearrangement)

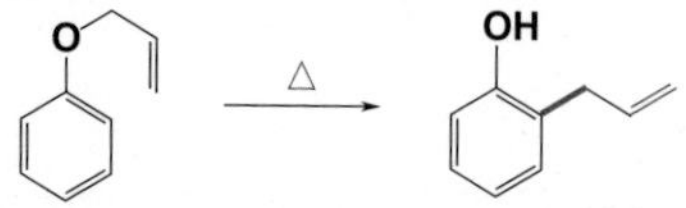

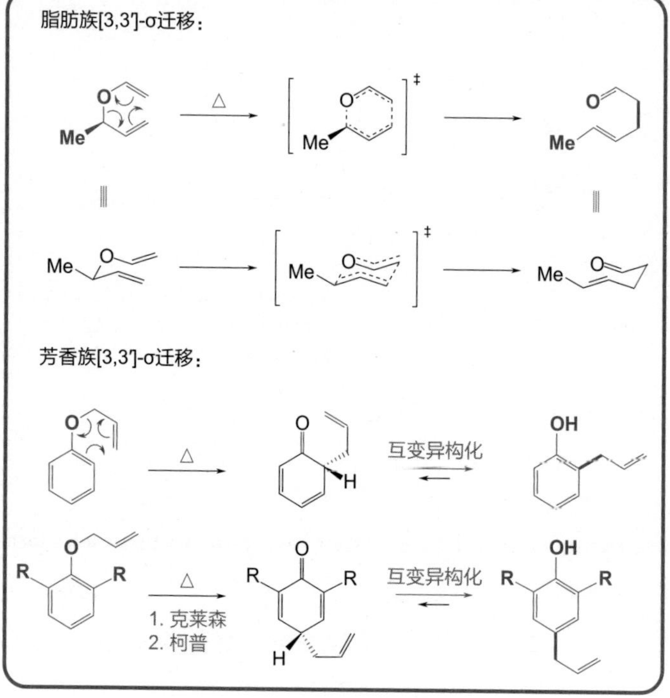

图 26.1 克莱森重排的机理 ❶

---

❶[译者注：克莱森重排是指烯醇类或酚类的烯丙基醚在加热条件下发生分子内重排，生成 γ, δ- 不饱和醛（酮）或邻（对）位烯丙基酚的反应。]克莱森重排与克莱森缩合（Claisen condensation）反应不同，但与柯普重排（Cope rearrangement）相似（参见第 28 个反应机理），是一种具有协同机制的周环反应。图中显示的是经典的 [3,3']-σ 重排，也可写作 [3,3']-σ 迁移。

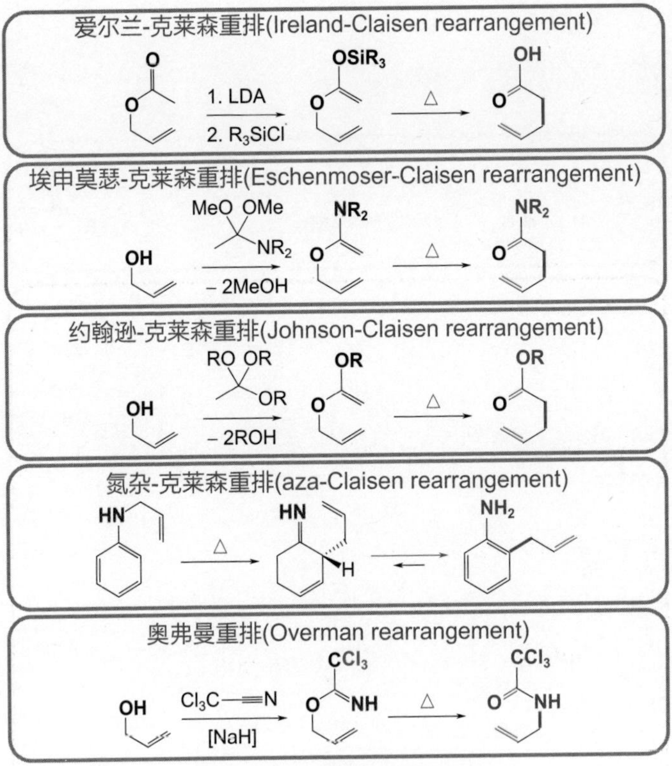

爱尔兰-克莱森重排(Ireland-Claisen rearrangement)
O
1. LDA
2. R_3SiCl
$OSiR_3$
O
△
OH
O
埃申莫瑟-克莱森重排(Eschenmoser-Claisen rearrangement)
MeO OMe
NR_2
OH
– 2MeOH
NR_2
O
△
NR_2
O
约翰逊-克莱森重排(Johnson-Claisen rearrangement)
RO OR
OR
OH
– 2ROH
OR
O
△
OR
O
氮杂-克莱森重排(aza-Claisen rearrangement)
HN
△
HN
H
NH_2
奥弗曼重排(Overman rearrangement)
OH
Cl_3C—≡N
[NaH]
CCl_3
O
NH
△
CCl_3
O
NH

图 26.2　**克莱森重排的相关反应**❶

1912
1800
1850
1900
1950
2000

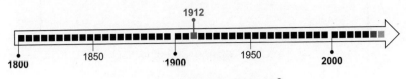

图 26.3　**克莱森重排反应的发现**❷

---

❶ 克莱森重排反应包括很多的变体和拓展反应：爱尔兰 - 克莱森重排、埃申莫瑟 - 克莱森重排、约翰逊 - 克莱森重排、氮杂 - 克莱森重排、氮杂 - 柯普重排和奥弗曼重排等[26a]。

❷ 该反应大约于 1912 年被首次报道[26b]。

27 柯普消除
(Cope elimination)

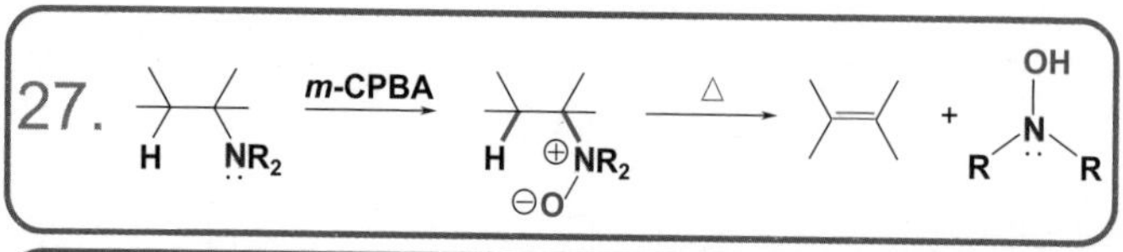

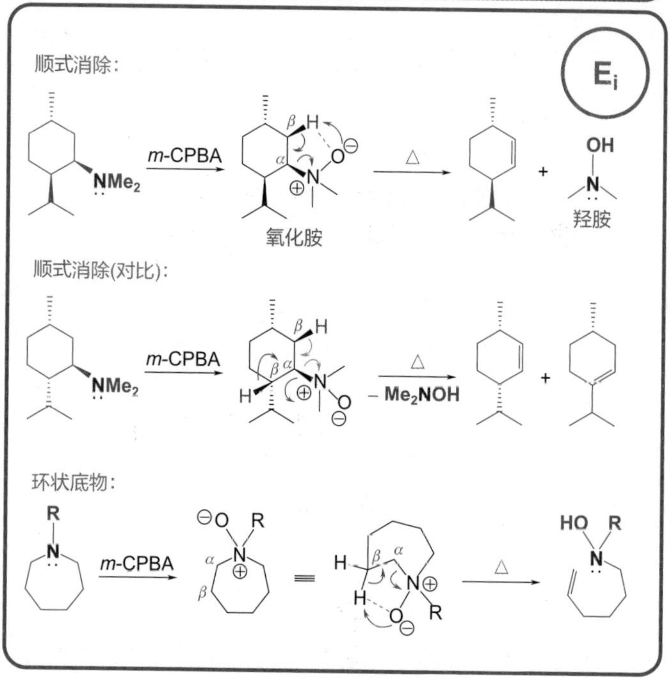

图 27.1 **柯普消除反应的机理** ❶

---

❶ 柯普消除，又称为柯普反应（Cope reaction）[译者注：是指 β-碳上含有氢的氧化胺在 150～200℃下发生消除反应生成羟胺和烯烃的反应]。该反应是典型的分子内或分子间 β-消除反应（E_i）（参见第 6 个反应机理）。

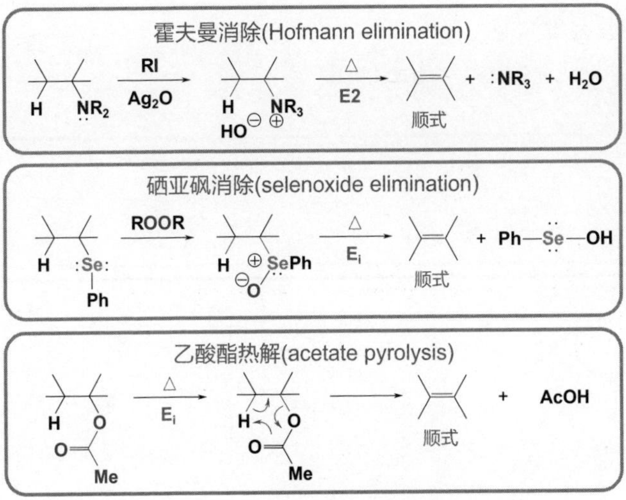

图 27.2 柯普消除的相关反应 ❶

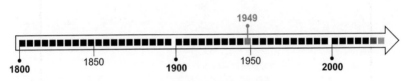

图 27.3 柯普消除反应的发现 ❷

---

❶ 与柯普消除相关的反应：霍夫曼消除 [通常是 E_2 消除，少部分情况是 E_1 消除（参见第 49 个反应机理）]、硒亚砜消除[27a, 27b]、乙酸酯热解[1]等。

❷ 该反应大约于 1949 年被首次报道[27c]。

28 柯普重排
(Cope rearrangement)

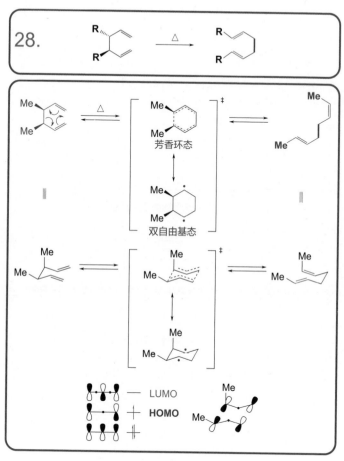

图 28.1　**柯普重排反应的机理** ❶

---

❶［译者注：柯普重排是指1,5-二烯的［3,3］-σ-迁移重排。］该反应与柯普消除反应不同，但与克莱森重排更为相似（参见第26个反应机理），是一种拥有特定机理的周环反应。图中显示的是经典的［3,3′］-σ重排（迁移）。

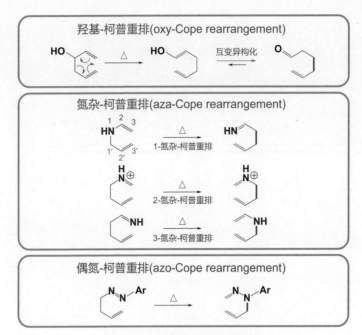

图 28.2 柯普重排的相关反应 ❶

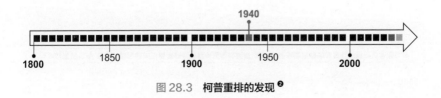

图 28.3 柯普重排的发现 ❷

---

❶ 柯普重排的相关反应较多[1]：(阴离子)羟基-柯普重排、氮杂-柯普/氮杂-克莱森重排、偶氮-柯普重排等[28a]。

❷ 该反应大约于1940年被首次报道[28b]。

29 克里奇氧化和马拉普拉德氧化
(Criegee & Malaprade oxidation)

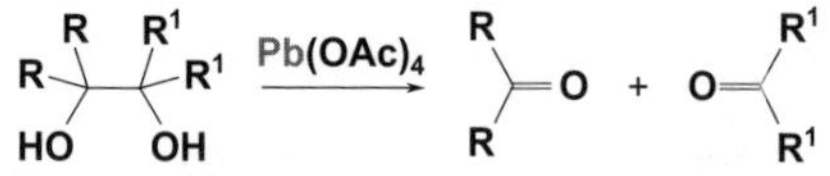

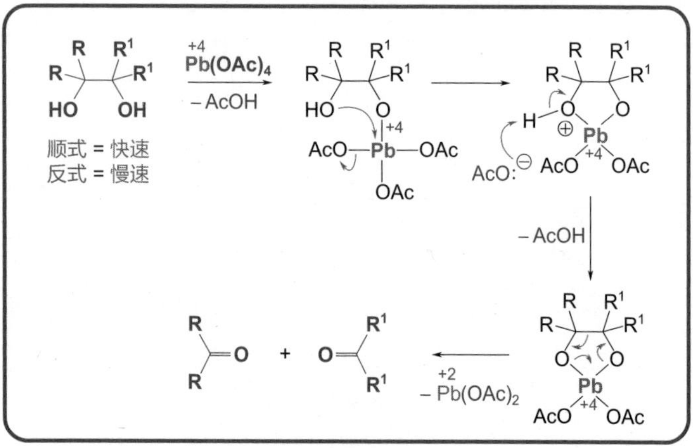

图 29.1 **克里奇氧化反应的机理** ❶

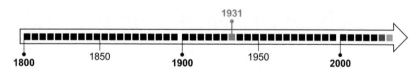

图 29.2 **克里奇氧化反应的发现** ❷

---

❶ 克里奇氧化（Criegee oxidation），又称为克里奇反应（Criegee reaction）[译者注：是指邻二醇被四乙酸铅氧化，经过环状酯中间体，邻二醇的碳-碳键断裂，醇羟基转化为相应的醛、酮的反应]。四乙酸铅参与的克里奇反应与臭氧参与的克里奇反应在机理上有所不同（参见第 70 个反应机理）。

❷ 该反应大约于 1931 年被首次报道[29a]。

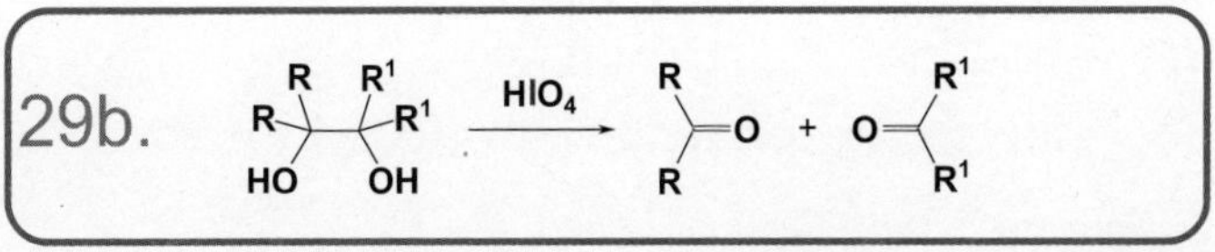

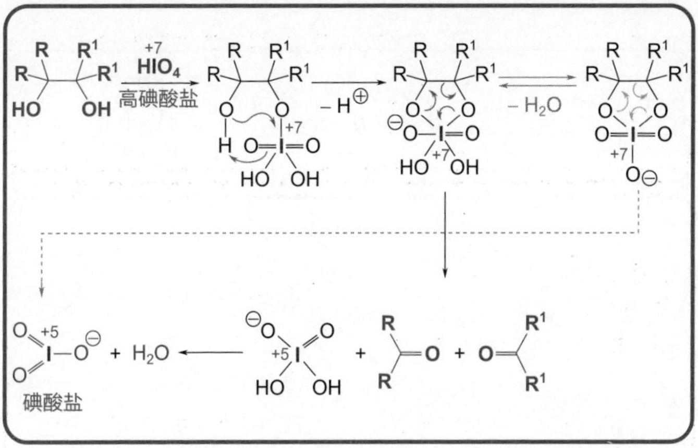

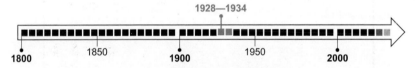

图 29.3　**马拉普拉德氧化（Malaprade oxidation）反应的机理** ❶

图 29.4　**马拉普拉德氧化反应的发现** ❷

---

❶ 马拉普拉德氧化与格里克反应相类似。
❷ 该反应大约于 1928—1934 年被首次报道[29b, 29c]。

30 铜催化叠氮-炔环加成
(copper-catalyzed azide-alkyne cycloaddition, CuAAC)

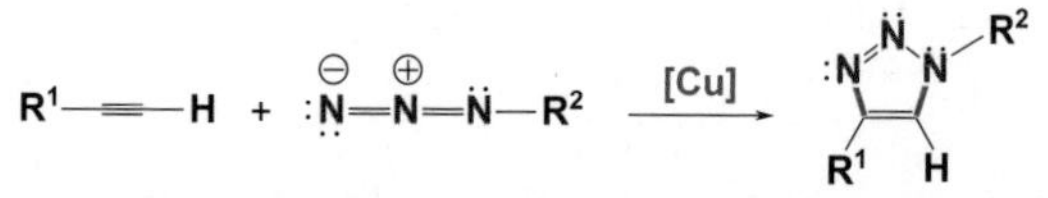

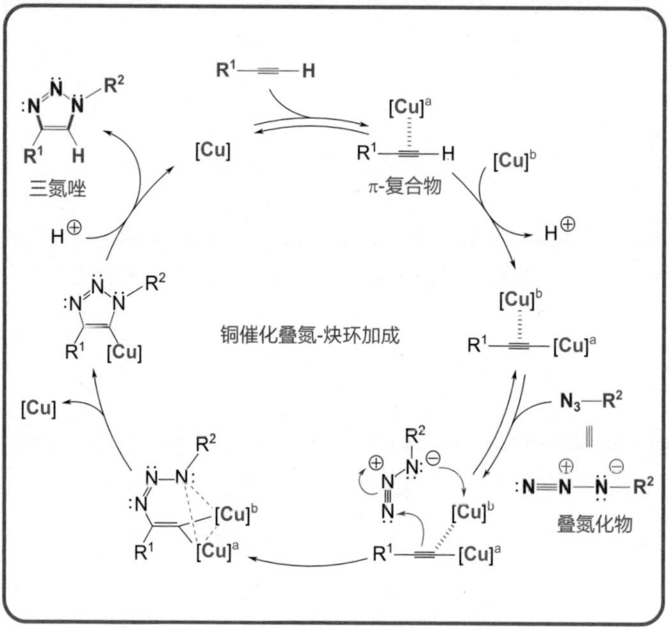

图 30.1　铜催化叠氮-炔环加成反应的机理 ❶

---

❶ [译者注：铜催化叠氮-炔环加成反应是指经 Cu（Ⅰ）催化，炔基与叠氮基发生环加成生成区域选择性的 1,4-二取代-1,2,3-三氮唑的反应。] CuAAC 是"点击化学"（click chemistry）的经典反应。具体而言，该反应是一种 1,3-偶极环加成或（3+2）-环加成反应。注意，（3+2）中的数字表示参与反应的原子个数，而 [4+2] 表示参与反应的电子个数[30a]。国际纯粹与应用化学联合会（IUPAC）并不建议将这两者混用，但在文献中 [3+2] 的使用频率更高。

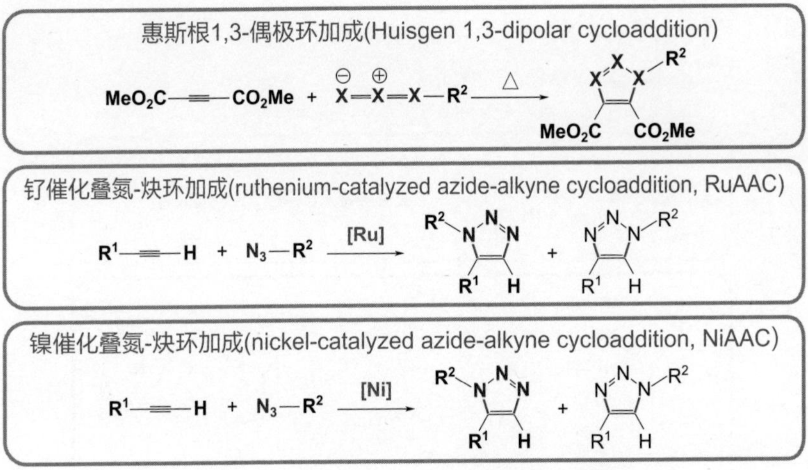

惠斯根1,3-偶极环加成(Huisgen 1,3-dipolar cycloaddition)

$MeO_2C{-}{\equiv}{-}CO_2Me + \overset{\ominus}{X}{=}X{=}\overset{\oplus}{X}{-}R^2 \xrightarrow{\Delta}$ 产物

钌催化叠氮-炔环加成(ruthenium-catalyzed azide-alkyne cycloaddition, RuAAC)

$R^1{-}{\equiv}{-}H + N_3{-}R^2 \xrightarrow{[Ru]}$ 产物

镍催化叠氮-炔环加成(nickel-catalyzed azide-alkyne cycloaddition, NiAAC)

$R^1{-}{\equiv}{-}H + N_3{-}R^2 \xrightarrow{[Ni]}$ 产物

图 30.2　铜催化叠氮–炔环加成的相关反应 ❶

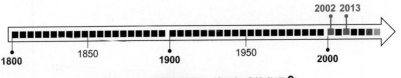

图 30.3　铜催化叠氮–炔环加成的发现 ❷

---

❶ 惠斯根环加成[30b, 30c]并不需要催化，但仍与 CuAAC 相关。叠氮化物 - 炔烃环加成也可以由钌（RuAAC）或镍（NiAAC）催化，但其机理与铜催化完全不同。

❷ 该反应大约于 2002 年被首次报道[30d, 30e]，而其反应机理于 2013 年被报道[30f]。

31 柯提斯重排
(Curtius rearrangement)

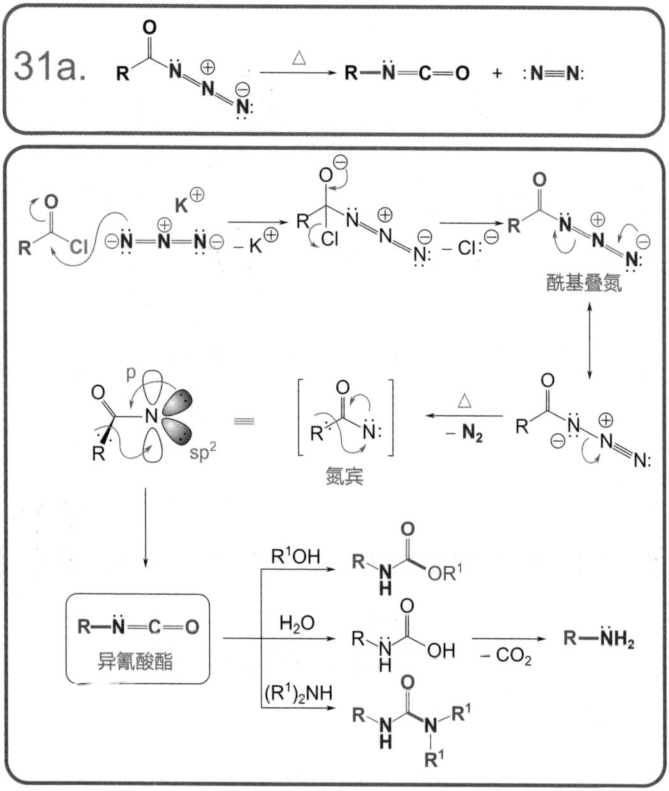

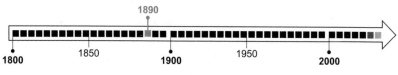

图 31.1　柯提斯重排的机理❶

图 31.2　柯提斯重排的发现❷

---

❶ 柯提斯重排，又称为柯提斯反应（Curtius reaction）[译者注：是指羧酸与叠氮化物作用生成酰基叠氮化物，再重排为异氰酸酯。异氰酸酯水解可生成少一碳的伯胺，该反应可用于几乎所有羧酸，是制备伯胺的重要方法]。

❷ 柯提斯重排大约于 1890 年被首次报道[31a, 31b]。

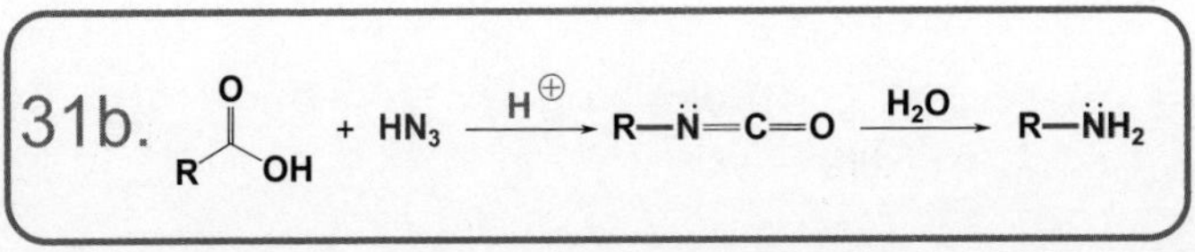

31b.

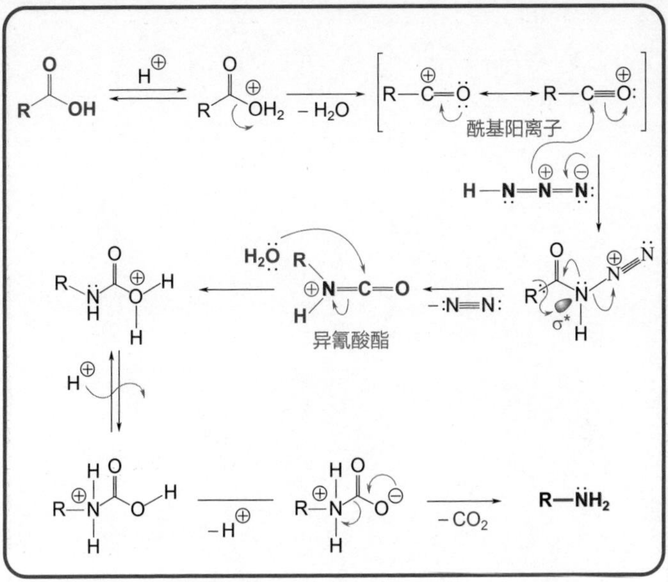

$$R\text{-}COOH + HN_3 \xrightarrow{H^{\oplus}} R\text{-}\ddot{N}\text{=}C\text{=}O \xrightarrow{H_2O} R\text{-}\ddot{N}H_2$$

图 31.3 施密特反应（Schmidt reaction）的机理 ❶

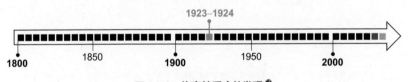

图 31.4 施密特反应的发现 ❷

---

❶ 施密特反应也是一种由羧酸制备伯胺的重排反应。
❷ 该反应大约于 1923—1924 年被首次报道[31c, 31d]。

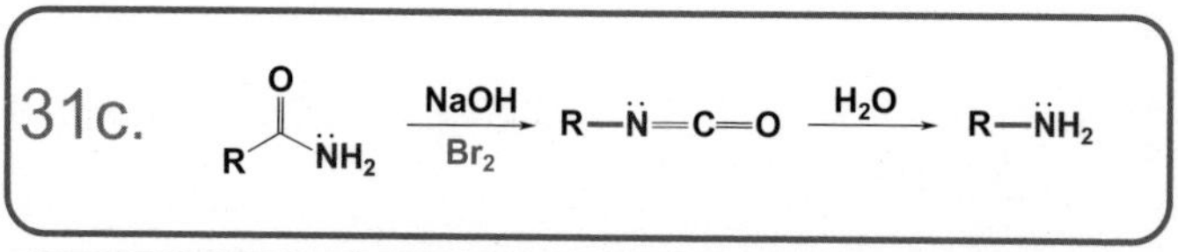

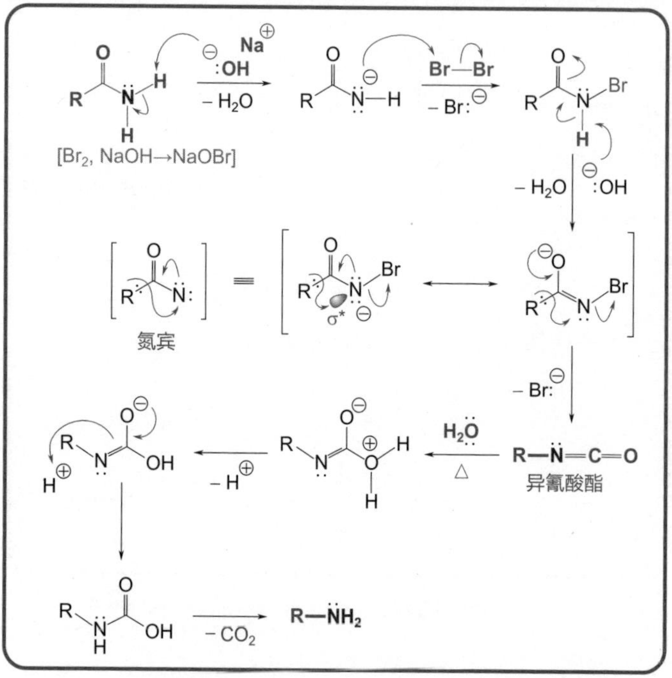

31c.

$$R-\underset{\underset{NH_2}{|}}{\overset{O}{C}} \xrightarrow{\underset{Br_2}{NaOH}} R-\ddot{N}=C=O \xrightarrow{H_2O} R-\ddot{N}H_2$$

图 31.5　霍夫曼重排（Hofmann rearrangement）的机理 ❶

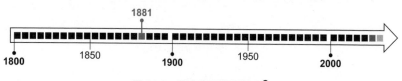

图 31.6　霍夫曼重排的发现 ❷

---

❶ 霍夫曼重排又称为霍夫曼反应 [译者注：是指伯酰胺重排转变为减少一个碳原子的伯胺的反应]。霍夫曼重排与霍夫曼消除的机理完全不同（参见第 49 个反应机理）。

❷ 霍夫曼重排大约于 1881 年被发现[31e]。

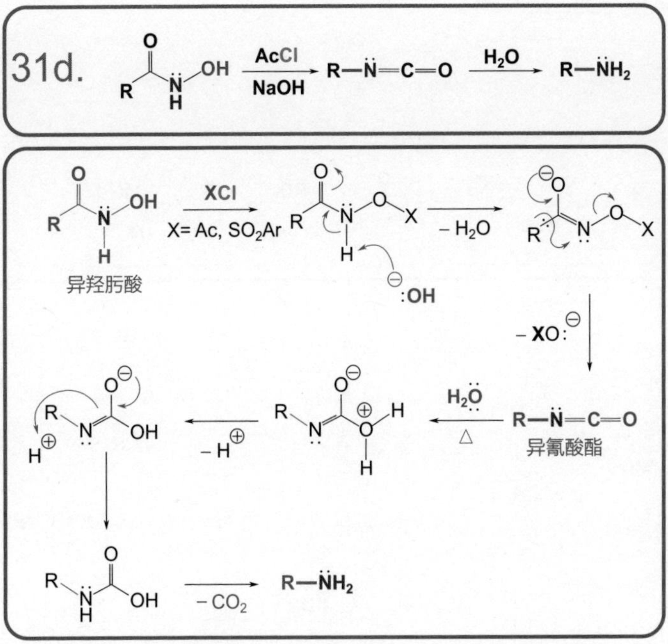

图 31.7 洛森重排（Lossen rearrangement）的机理 ❶

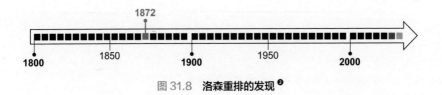

图 31.8 洛森重排的发现 ❷

---

❶ [译者注：洛森重排是指异羟肟酸衍生物经加热或碱作用后，经由异氰酸酯中间体最终生成伯胺的反应。] 洛森重排与贝克曼重排相关（参见第 14 个反应机理）。

❷ 该反应大约于 1872 年被首次报道[31f]。

32 达森缩合
(Darzens condensation)

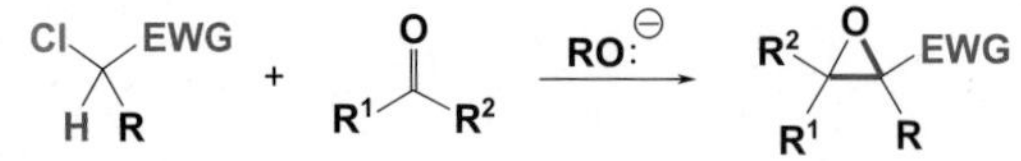

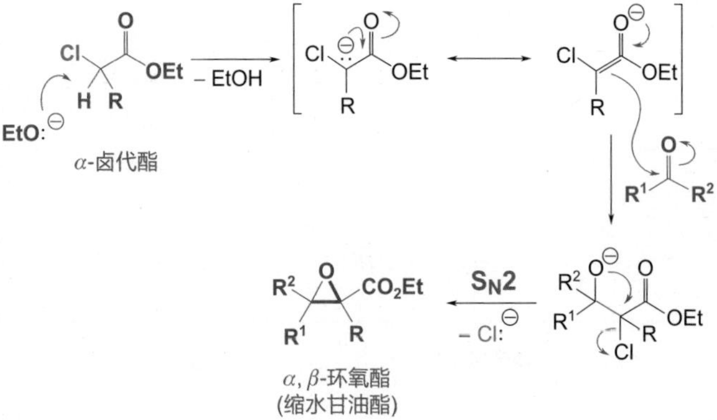

图 32.1 达森斯缩合反应的机理 ❶

---

❶ 达森斯缩合，又称为达森反应（Darzens reaction）或达森缩水甘油酯缩合（Darzens glycidic ester condensation）[译者注：是指醛或酮在强碱（如氨基钠、醇钠）作用下与 α-卤代酯反应生成 α, β-环氧酯的反应]。

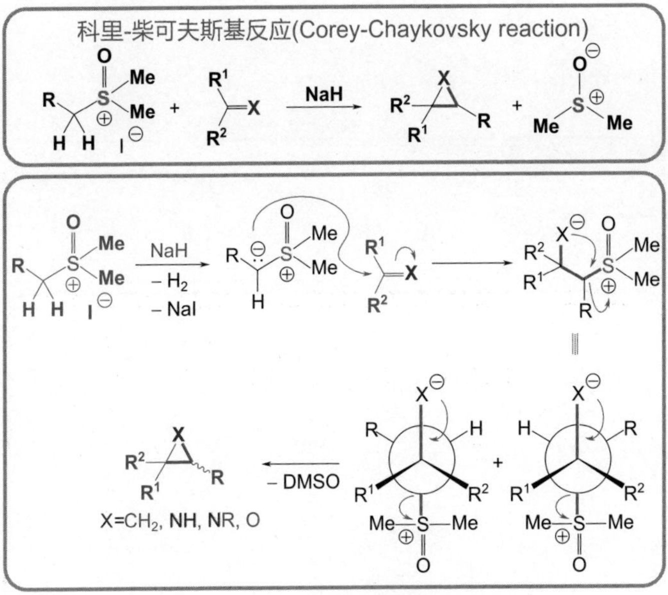

32 达森缩合

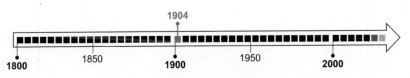

图 32.2 科里－柴可夫斯基反应的机理 ❶

图 32.3 达森缩合反应的发现 ❷

---

❶ 科里-柴可夫斯基反应，又称为约翰逊-科里-柴可夫斯基反应（Johnson-Corey-Chaykovsky reaction）[32a, 32b]，与达森缩合（Darzens condensation）反应及维蒂希反应（Wittig reaction）（参见第 98 个反应机理）有关联。

❷ 该反应大约于 1904 年被首次报道[32c]。

33 戴斯-马丁氧化
(Dess-Martin oxidation)

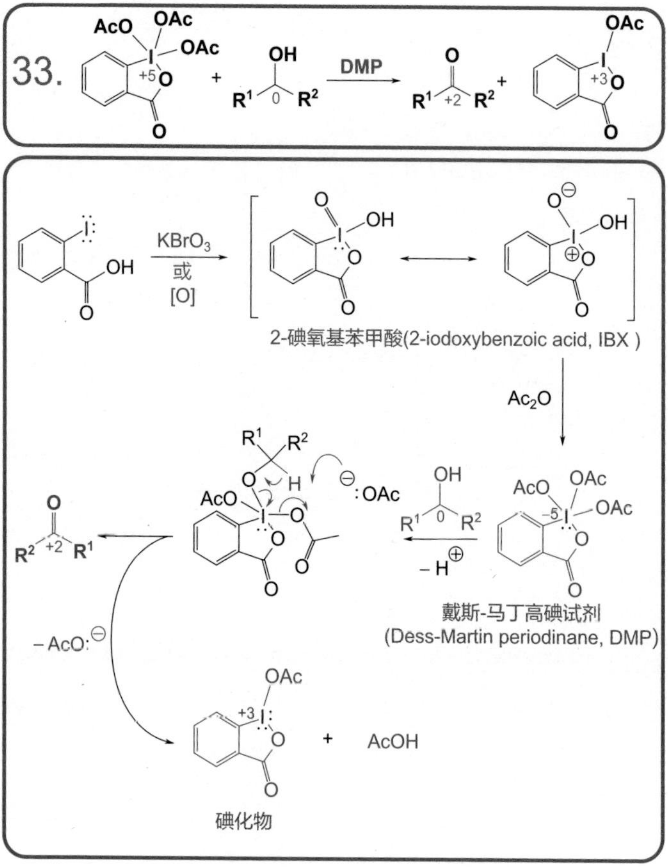

图 33.1 **戴斯-马丁氧化反应的机理**[1]

---

[1] 戴斯-马丁氧化反应是依据同名的戴斯-马丁氧化剂命名的,是指通过该氧化剂将醇氧化成醛和酮的反应[33a, 33b]。

33 戴斯-马丁氧化

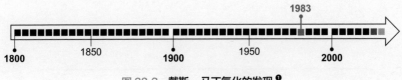

图 33.2　戴斯－马丁氧化的发现 ❶

---

❶ 该反应大约于 1983 年被首次报道[33c]。

34 重氮化反应
(diazotization)

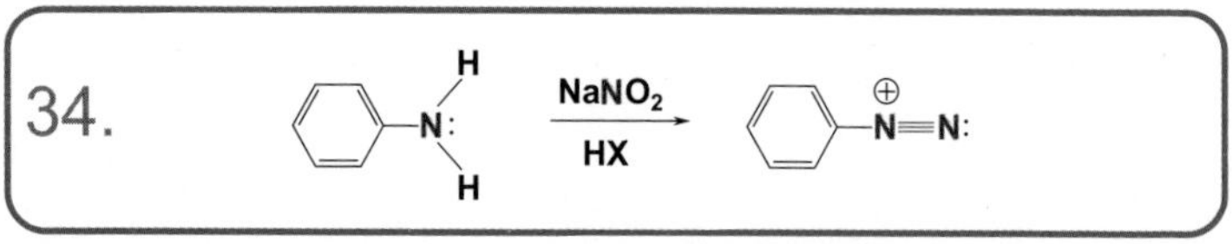

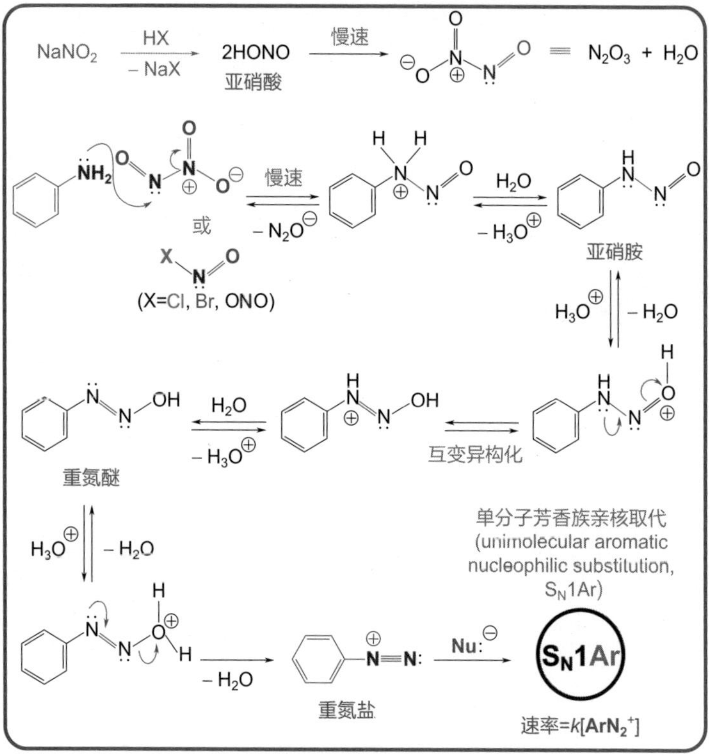

图 34.1 **重氮化（重氮盐）反应的机理** ❶

---

❶ 重氮化（diazotization[1]，diazoniation[1a] 或 diazotation[34a]）反应，又称为重氮盐（diazonium salt）反应，是指芳香族伯胺和亚硝酸作用（在强酸介质下）生成重氮盐的反应。

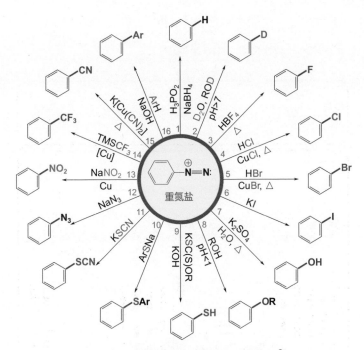

图 34.2 重氮盐在有机合成中的广泛应用 ❶

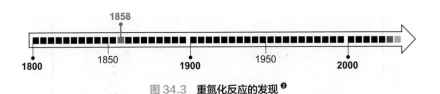

图 34.3 重氮化反应的发现 ❷

---

❶ 重氮化反应生成的重氮盐具有广泛的合成应用,可以与多种亲核试剂反应。该反应的机理为单分子芳香亲核取代(unimolecular aromatic nucleophilic substitution,S_N1Ar)或自由基链式亲核取代反应(radical chain nucleophilic substitution reaction,$S_{RN}1$)。S_N1Ar 表示单分子反应(unimolecular,Uni- 是 1 的意思),即反应速率为一级,并且限速步骤的速率只取决于重氮盐底物(ArN_2^+)的浓度:速率$= k\,[ArN_2^+]$。该反应的机理与第 4 个反应机理中介绍的加成消除机理(S_NAr 或 S_N2Ar)不同,因为反应第一步为消除步骤并形成芳基阳离子。注意,该反应也与第 16 个反应机理中介绍的苯炔机理(消除加成机理)不同。

❷ 该反应大约于 1858 年被首次报道[34b]。

35 狄尔斯-阿尔德环加成
(Diels-Alder cycloaddition)

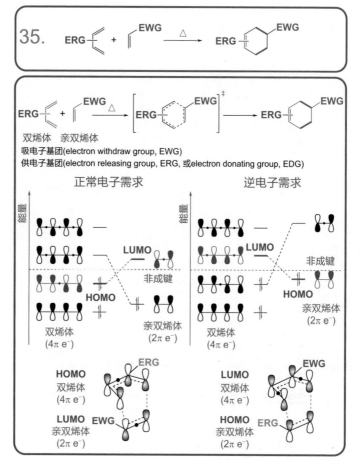

图 35.1 狄尔斯-阿尔德环加成反应的机理 ❶

---

❶ 狄尔斯-阿尔德环加成，又称为狄尔斯-阿尔德反应（Diels-Alder reaction）或[4+2]-环加成反应（[4+2]-cycloaddition reaction），具体指共轭双烯与取代烯烃（一般称为亲双烯体）生成取代环己烯的反应，是一类具有特定机理的周环反应。注意，(4+2)中的数字表示参与反应的原子个数，而[4+2]表示参与反应的电子个数[30a]。可与1,3-偶极环加成反应相类比（参见第30个反应机理）。

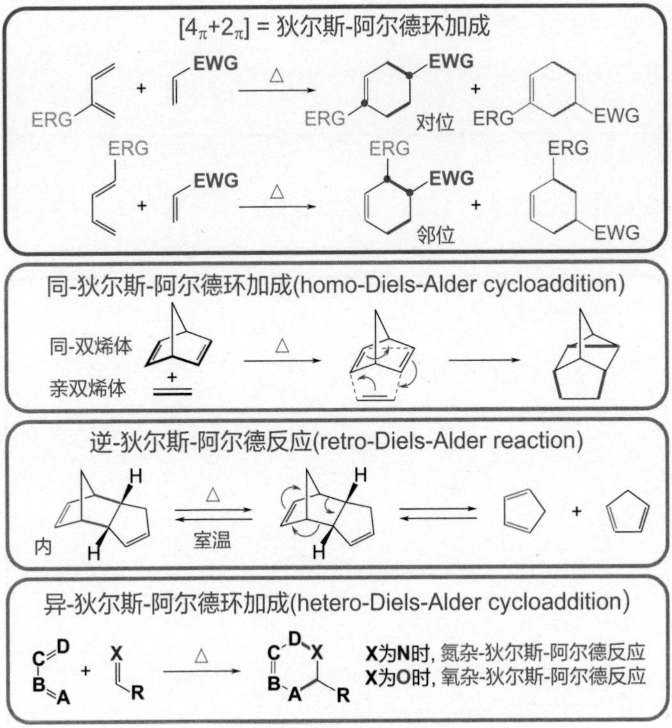

图 35.2 狄尔斯－阿尔德环加成的相关反应 ❶

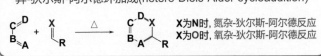

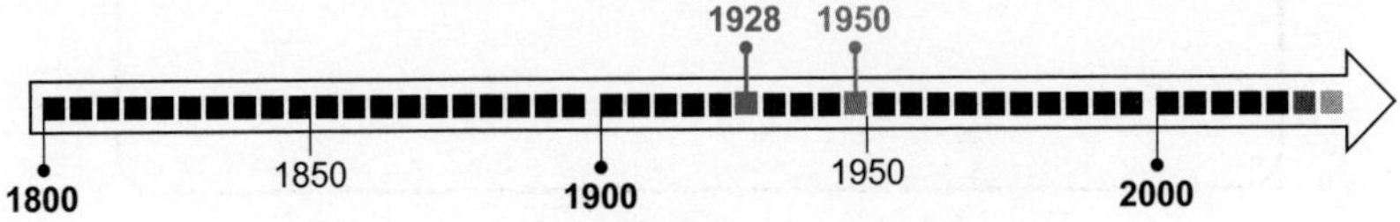

图 35.3 狄尔斯－阿尔德环加成反应的发现 ❷

---

❶ 狄尔斯 - 阿尔德环加成有很多变体反应：同 - 狄尔斯 - 阿尔德环加成、逆 - 狄尔斯 - 阿尔德反应、异 - 狄尔斯 - 阿尔德环加成等。第一个实例 [$4_\pi+2_\pi$] = Diels-Alder 环加成中的区域化学性需要予以注意。

❷ 狄尔斯 - 阿尔德环加成大约于 1928 年被首次报道[33a, 33b]。Otto Paul Hermann Diels 和 Kurt Alder 因发现这一双烯合成方法而获得了 1950 年的诺贝尔化学奖[35c]。

36 双-π-甲烷重排
(di-π-methane rearrangement)

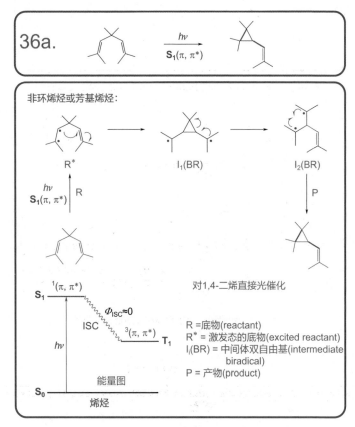

图 36.1 双-π-甲烷重排的机理：直接光催化 ❶

---

❶ 双-π-甲烷重排（DPM）在特定情况下又被称为齐默尔曼反应（Zimmerman reaction），是指1,4-二烯光解生成乙烯基环丙烷的反应。如果反应被光直接催化，该反应将从单重激发态 S_1 开始反应，图中为 $^1(\pi, \pi^*)$[2b]。

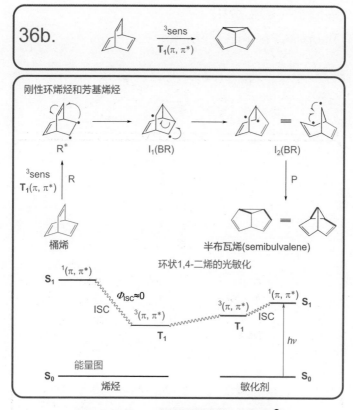

图 36.2　双 -π- 甲烷重排的机理：光敏化 ❶

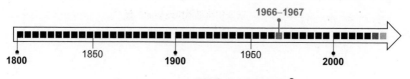

图 36.3　双 -π- 甲烷重排反应的发现 ❷

---

❶ 在光敏剂的存在下，双 -π- 甲烷重排将经历光敏化；产物将从三重激发态 T_1 开始反应，图中为 $^3(\pi, \pi^*)$[2b]。

❷ 该反应大约于 1966—1967 年被首次报道[36]。

37 法沃尔斯基重排
(Favorskii rearrangement)

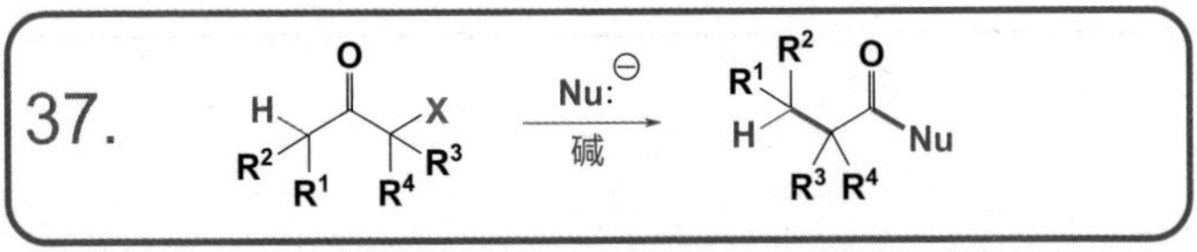

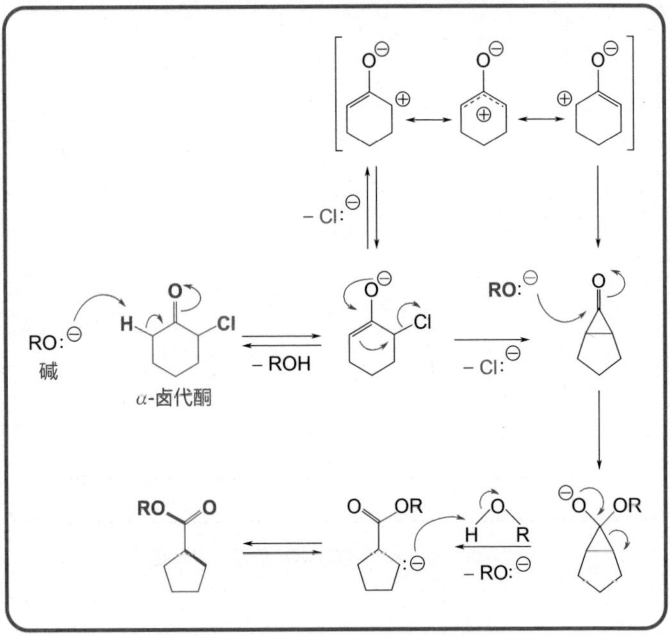

图 37.1 法沃尔斯基重排反应的机理 ❶

---

❶ 法沃尔斯基重排（Favorskii rearrangement，Favorsky rearrangement）是指在醇钠、氢氧化钠、氨基钠等碱性催化剂存在下，α-卤代酮（α-氯代酮或α-溴代酮）失去卤离子，重排成具有少一个碳原子的羧酸酯、羧酸、酰胺的反应。该反应与法沃尔斯基反应（Favorskii reaction）完全不同。

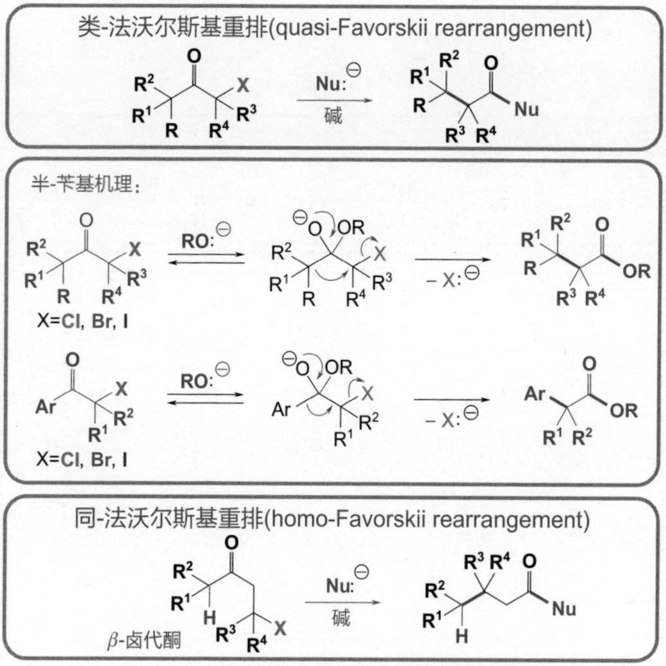

图 37.2 类-法沃尔斯基重排的机理及其相关反应 ❶

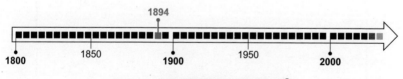

图 37.3 法沃尔斯基重排反应的发现 ❷

---

❶ 法沃尔斯基基重排（Favorskii rearrangement）包括多种变体反应：类-法沃尔斯基重排（机理与半-苄基机理类似[37a, 37b]），同-法沃尔斯基重排等反应。

❷ 该反应大约于1894年被首次报道[37c, 37d]。

38 费歇尔吲哚合成
(Fischer indole synthesis)

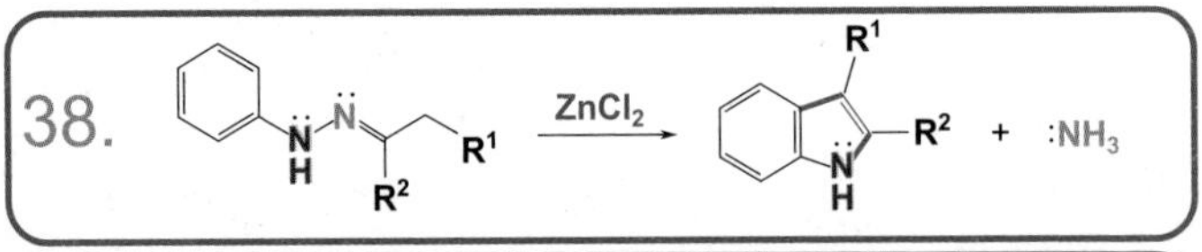

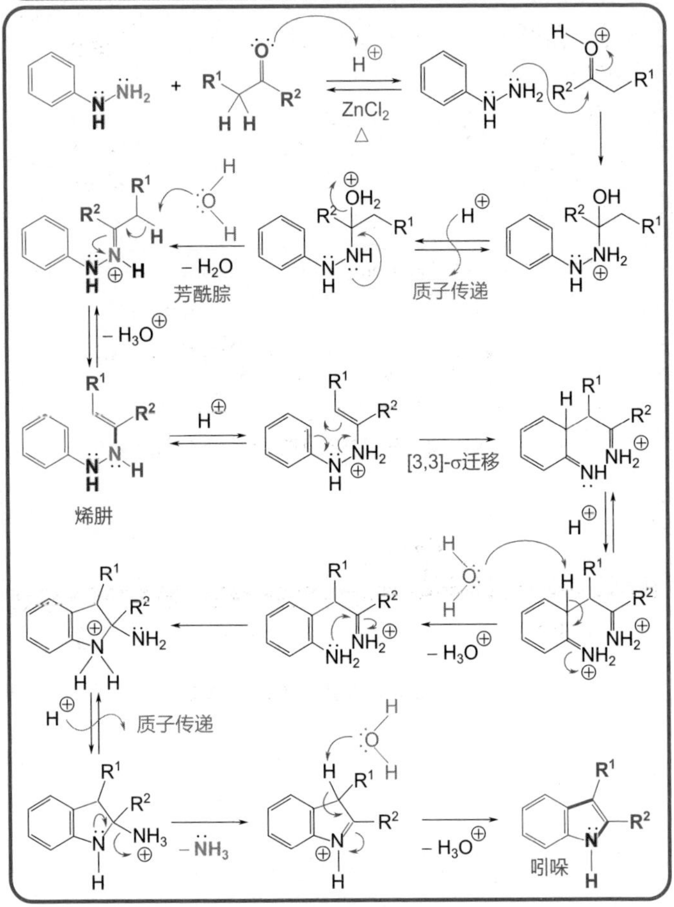

图 38.1 费歇尔吲哚合成反应的机理 ❶

---

❶ [译者注：费歇尔吲哚合成是合成吲哚的常用方法，具体是以苯肼与醛、酮在酸催化下加热重排消除一分子氨，得到 2- 或 3- 取代的吲哚的反应。] 费歇尔吲哚合成与费歇尔酯化反应不同，它是有机化学中最重要的反应之一。其中的关键步骤是 [3,3']-σ 迁移 (重排)。

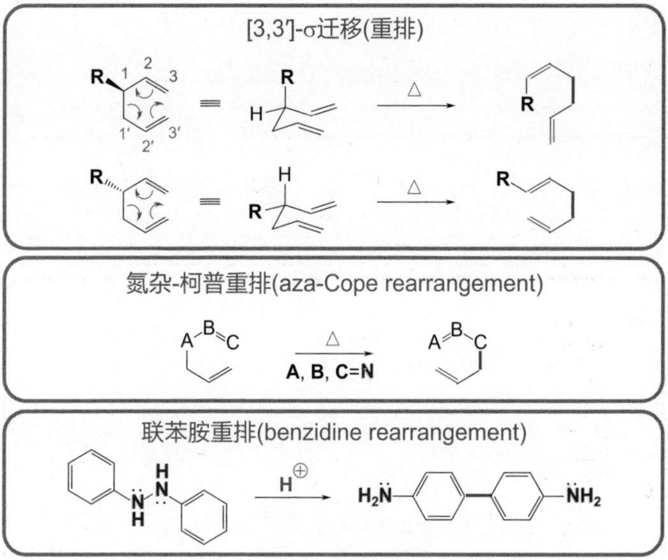

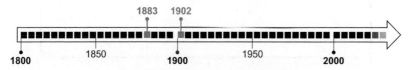

图 38.2　费歇尔吲哚合成的相关反应 ❶

图 38.3　费歇尔吲哚合成反应的发现时间 ❷

---

❶ 关键步骤与柯普重排、氮杂-柯普/氮杂-克莱森重排有关(参见第 28 个反应机理)。与该转化有关的其他反应包括联苯胺重排(其机理尚不清楚)[1, 38a]。

❷ 费歇尔吲哚合成大约于 1883 年被首次报道[38b, 38c]。1902 年,Emil Fischer 凭借在该反应领域的成就获得了诺贝尔化学奖[38d]。

39 傅-克酰基化和傅-克烷基化
(Friedel-Crafts acylation & alkylation)

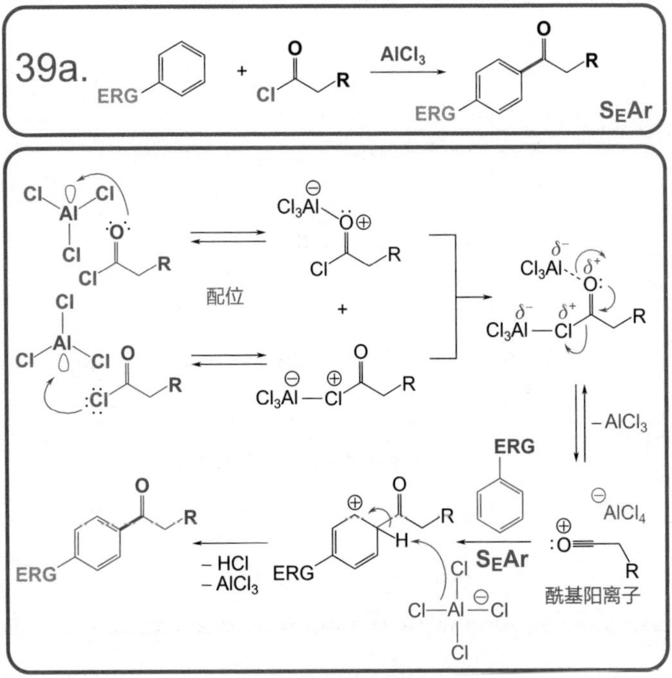

图 39.1 傅-克酰基化反应的机理 ❶

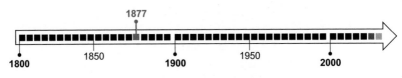

图 39.2 傅-克酰基化反应的发现 ❷

---

❶ [译者注：傅-克酰基化是指在路易斯酸存在的条件下，酰卤或酸酐与芳香化合物进行的酰化反应。] 傅-克酰基化机理属于芳香亲电取代 (aromatic electrophilic substitution，S_EAr) 的典型实例 (芳基正离子机理) (参见第 3 个反应机理)。链状酰卤通过酰基阳离子反应生成具有链状烷基链的芳基酮。

❷ 该反应大约于 1877 年被首次报道[39a]。

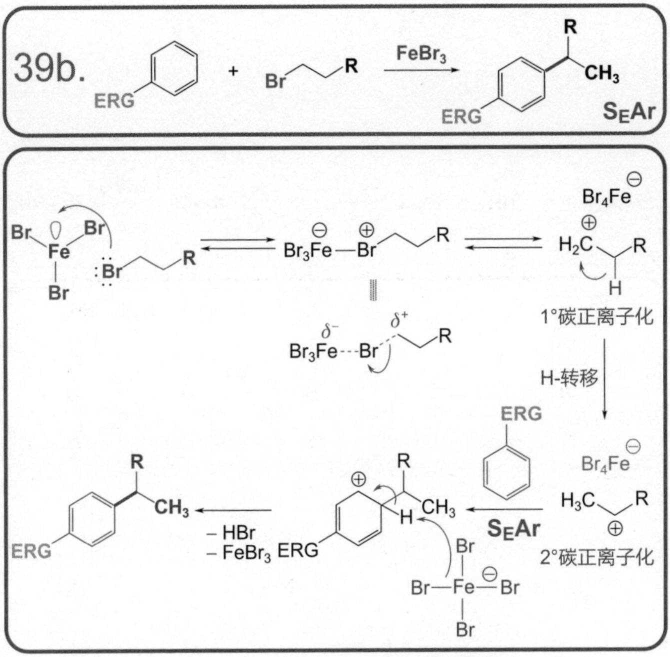

39 傅-克酰基化和傅-克烷基化

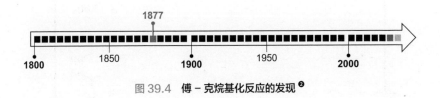

图 39.3　傅-克烷基化反应的机理 ❶

图 39.4　傅-克烷基化反应的发现 ❷

---

❶ 傅-克烷基化机理也属于芳香亲电取代。链状卤代烷经历碳正离子重排[又称为瓦格纳-麦尔外因重排(Wagner-Meerwein rearrangement)(参见第 96 个反应机理)]后更容易生成支链较多的产物。

❷ 该反应大约于 1877 年被首次报道[39b]。

40 盖布瑞尔合成
(Gabriel synthesis)

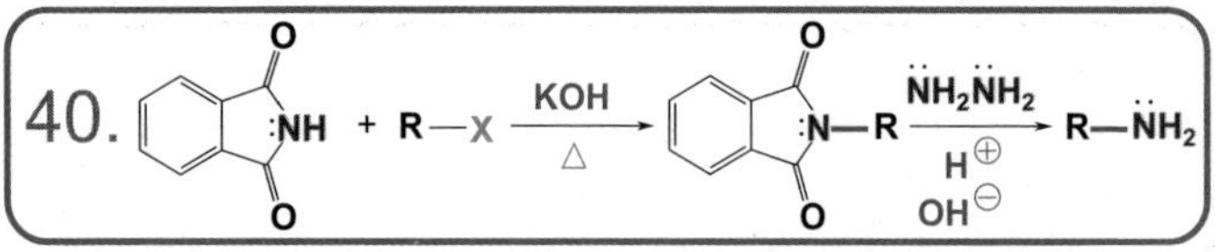

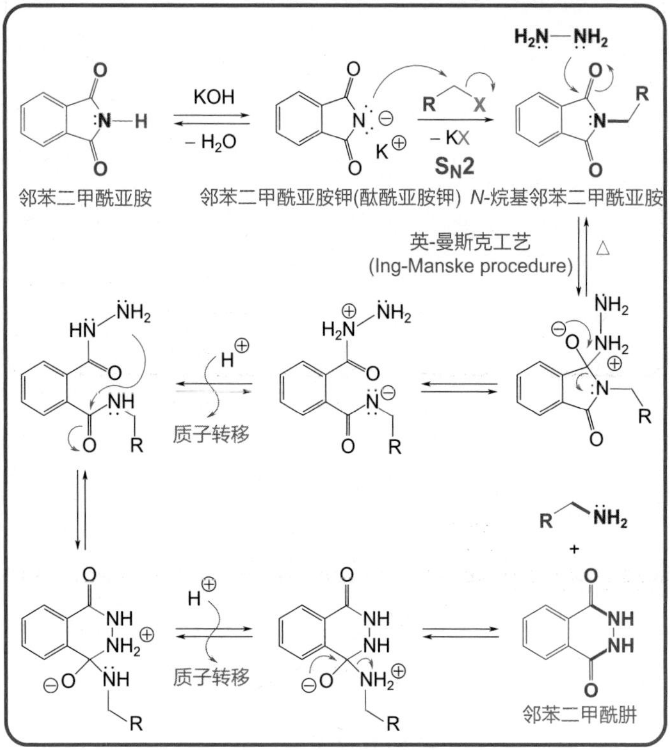

图 40.1　盖布瑞尔合成反应的机理 ❶

---

❶ [译者注：盖布瑞尔合成是指使用酰亚胺（琥珀酰亚胺、邻苯二甲酰亚胺）将卤代烷转化为伯胺的反应。] 盖布瑞尔合成属于 S_N2 反应。英 - 曼斯克工艺（Ing-Manske procedure）[40a]是一种使用肼将 N- 烷基邻苯二甲酰亚胺转化为伯胺的反应。

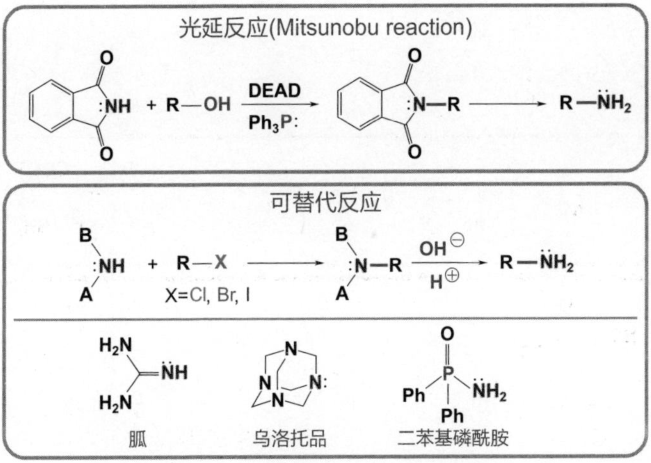

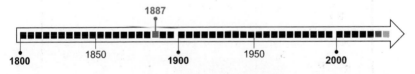

图40.2 盖布瑞尔合成的相关反应 ❶

图40.3 盖布瑞尔合成的发现 ❷

---

❶ 其他反应也可用于制备伯胺：光延反应（参见第61个反应机理），以及其他使用含氮亲核试剂的 S_N2 反应。其中一些也是人名反应：德尔宾反应（Delépine reaction），所用的含氮亲核试剂为乌洛托品[40b]。

❷ 该反应大约于1887年被首次报道[40c]。

41 格瓦尔德反应
(Gewald reaction)

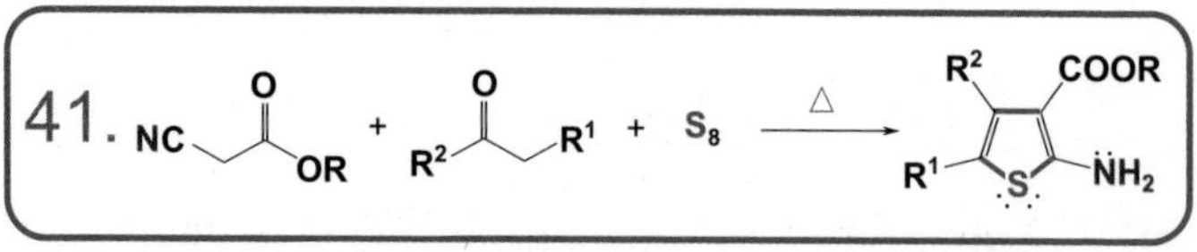

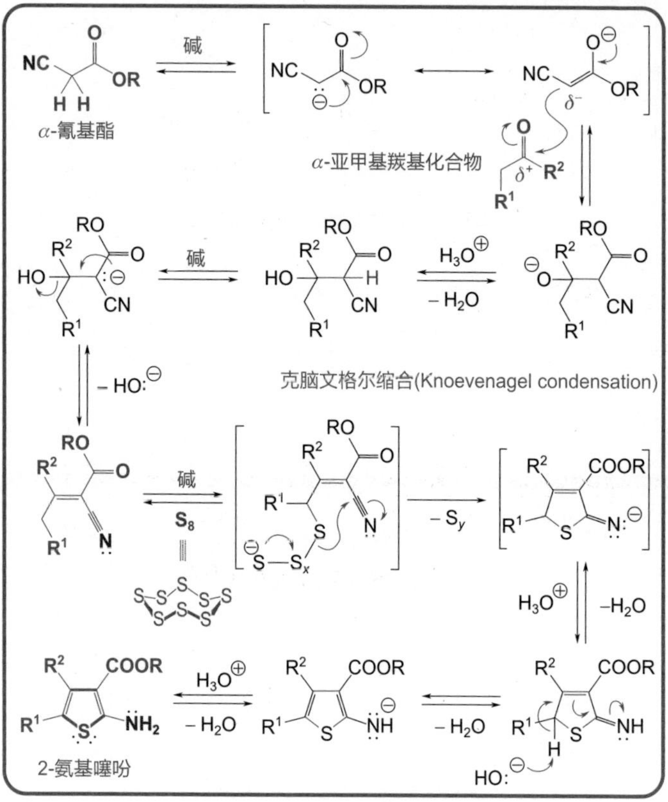

图 41.1 格瓦尔德反应的机理 ❶

---

❶ 格瓦尔德反应，又称为格瓦尔德缩合（Gewald condensation）[译者注：具体是指酮与 α-氰酯进行克脑文格尔缩合生成稳定的不饱和酯中间体，然后与硫进行加成，与碳相连的硫再进攻氰基进行关环，最终生成 2-氨基噻吩类化合物]。格瓦尔德反应属于三组分反应（3-component reaction, 3CR），关键的缩合步骤为克脑文格尔缩合[41a]。

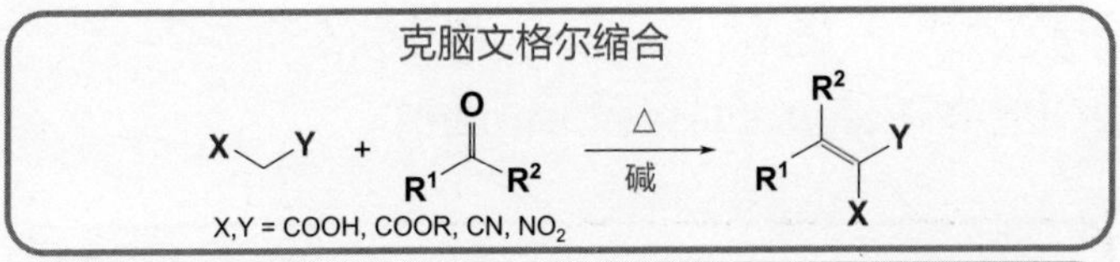

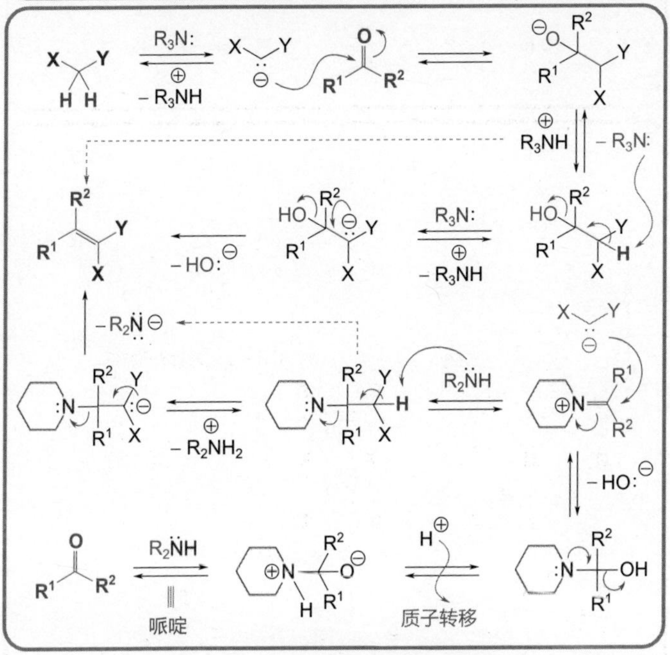

图 41.2 克脑文格尔缩合的机理 ❶

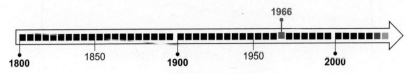

图 41.3 格瓦尔德反应的发现 ❷

---

❶[译者注：克脑文格尔缩合是指含有活泼亚甲基的化合物与醛或酮在弱碱催化下，发生失水缩合生成 α, β- 不饱和羰基化合物及其类似物的反应。]该反应是醇醛缩合的变体反应（参见第 83 个反应机理），通常由哌啶催化。

❷ 该反应大约于 1966 年被首次报道[41b]。

42 格拉泽 - 埃格林顿 - 海偶联
(Glaser-Eglinton-Hay coupling)

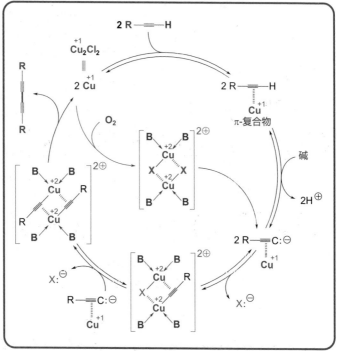

图 42.1 **格拉泽 - 埃格林顿 - 海偶联的机理** ❶

---

❶ 格拉泽 - 埃格林顿 - 海偶联是三个人名反应的通用名称：格拉泽偶联、埃格林顿偶联和海偶联。这些反应都属于铜催化末端炔烃的二聚化反应。在这三种反应中，产物都是对称的。

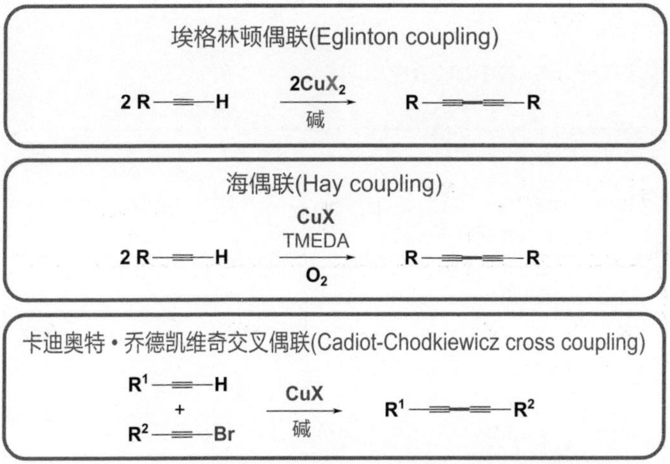

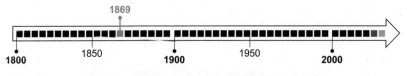

图 42.2 格拉泽 – 埃格林顿 – 海偶联的相关反应 ❶

图 42.3 格拉泽 – 埃格林顿 – 海偶联反应的发现 ❷

---

❶ 埃格林顿偶联反应具有如下特点：(a)产物是对称的；(b)铜被用作化学当量试剂[42a, 42b]。格拉泽偶联具有如下特点：(a)产物是对称的；(b) CuX 用作 NH_3 或 NH_4OH 的催化剂[42c]。海偶联反应具有如下特点：(a)产物是对称的；(b) CuX·TMEDA 复合物作为催化剂[42d, 42e]。卡迪奥特·乔德凯维奇交叉偶联反应具有如下特点：(a)产物是对称的；(b)铜作为催化剂[42f]。其他示例见文献[1,4]。

❷ 该反应大约于 1869 年被首次报道[42c]。

43 格氏反应
(Grignard reaction)

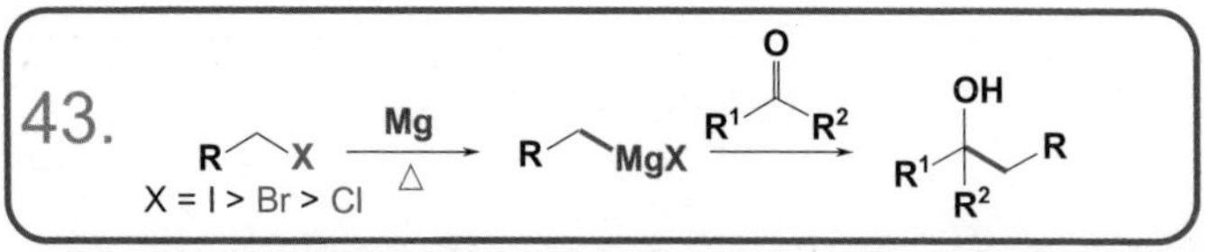

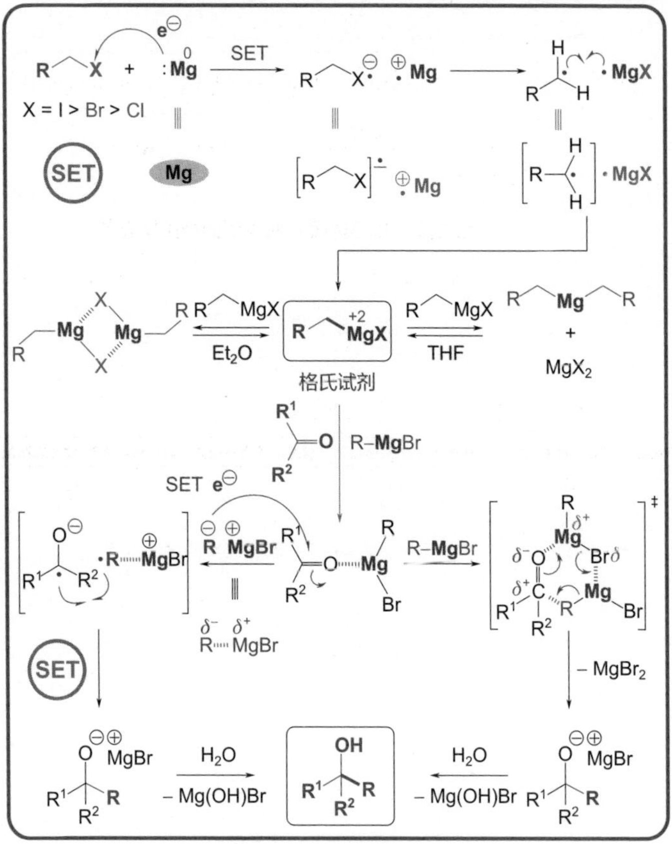

图 43.1　格氏反应的机理[1]

---

[1] 格氏反应的命名是基于同名试剂——格氏试剂（Grignard reagent）（RMgX）。[译者注：格氏反应是指卤代物在无水乙醚或四氢呋喃中和金属镁作用生成格氏试剂烷基卤化镁，然后格氏试剂作为亲核试剂与醛、酮、羧酸等化合物发生加成反应，构建C—C键。] 该机制尚未得到很好的阐释，可能涉及单电子转移（参见第5个反应机理）。

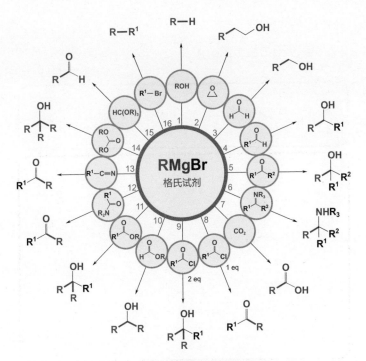

图 43.2　格氏试剂的合成多功能性 ❶

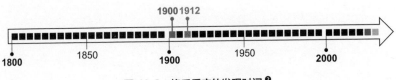

图 43.3　格氏反应的发现时间 ❷

---

❶ 格氏试剂在合成上具有广泛的应用，可与多种亲电试剂（亲电中心）反应：1. 醇、氘化水；2. 环氧化物；3. 甲醛；4. 其他醛类；5. 酮类；6. 亚胺类；7. 二氧化碳（二硫化物）；8. 酰氯（1当量）；9. 酰氯（过量）；10. 甲酸酯；11. 酯；12. 酰胺；13. 腈；14. 碳酸盐；15. 原酸酯；16. 卤代烷等[1]。

❷ 该反应大约于 1900 年被首次报道[43a]。鉴于 Victor Grignard 与 Paul Sabatier 在格氏试剂发现等领域的成就，二人于 1912 年获得了诺贝尔化学奖[43b]。

44 格罗布裂解
(Grob fragmentation)

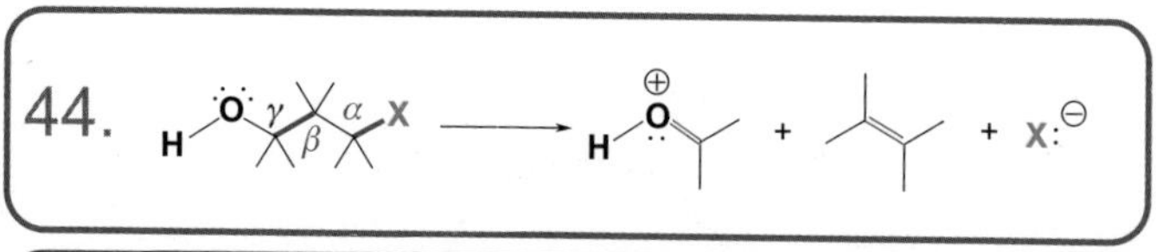

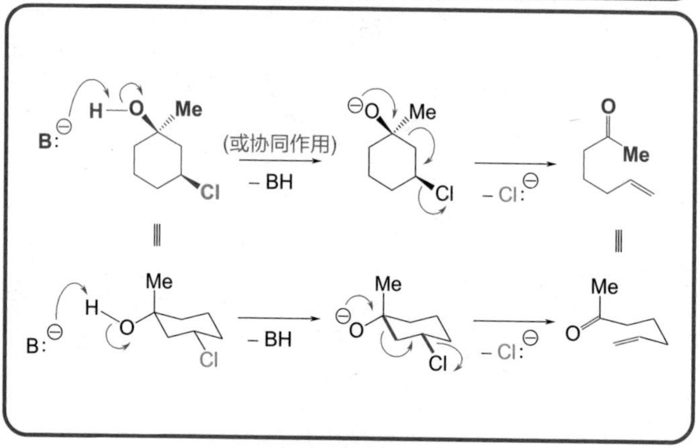

图 44.1 格罗布裂解反应的机理 ❶

---

❶[译者注：格罗布裂解反应是指脂肪链上 1 位和 3 位含有离电体（electrofuge，E）和离核体（nucleofuge，N）的化合物断裂成三个片段的反应。]该反应的机理最有可能与第 6 个反应机理所述的 β- 消除（在这种情况下为 1,4- 消除）有关。这种裂解反应的共同特征是形成了三种物质：带正电的片段（离电体）、中性的不饱和片段，以及带负电荷的片段（离核体）。在这一过程中包含多步骤的机理。

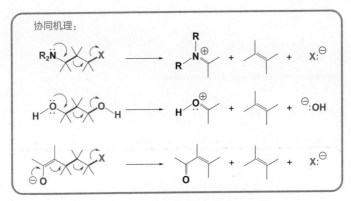

图44.2 格罗布裂解反应的相关反应 ❶

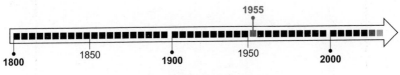

图44.3 格罗布裂解反应的发现 ❷

❶ 很多底物可发生格罗布裂解，如 γ- 羟基卤化物（γ-hydroxy halides）、γ- 氨基卤化物（γ-amino halides）、1,3- 二醇（1,3-diols）等[44a]。

❷ 该反应大约于 1955 年被首次报道[44b, 44c]。

45 卤仿反应
(haloform reaction)

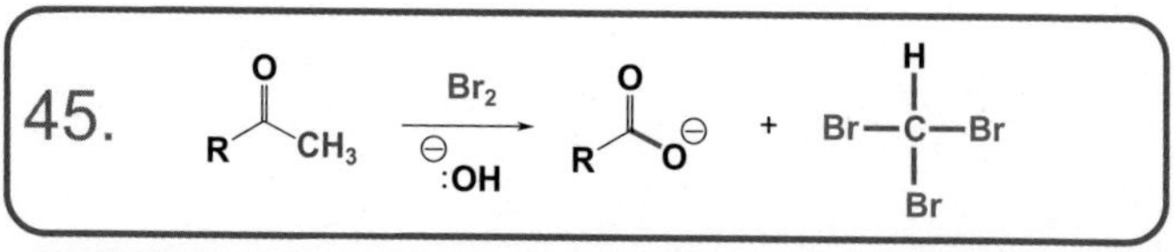

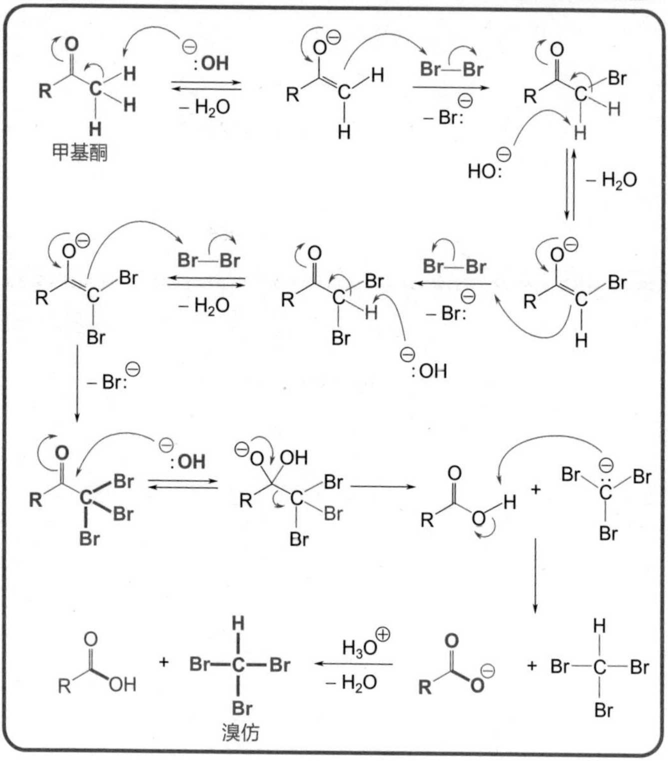

图 45.1　**卤仿反应的机理** ❶

---

❶ [译者注：卤仿反应是指甲基酮类化合物，即含有乙酰基的化合物在碱性条件下卤化并生成卤仿的反应。] 卤仿反应是有机化学中最古老的反应之一，属于脂肪族亲电取代的实例。

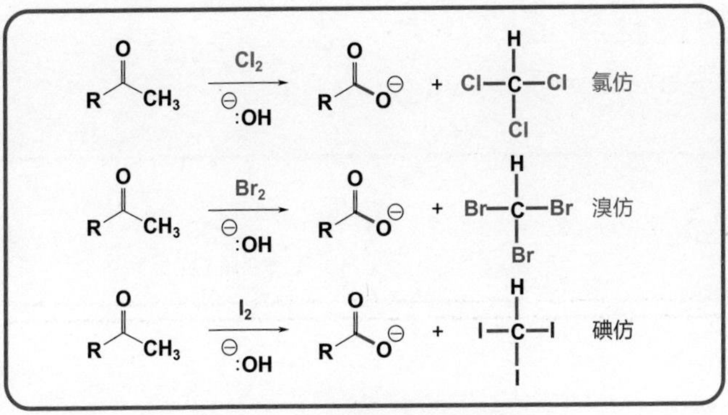

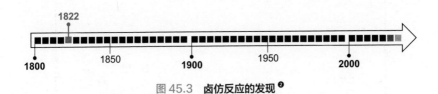

图 45.2　卤仿反应的相关反应 ❶

图 45.3　卤仿反应的发现 ❷

---

❶ 卤仿反应可用于大多数卤素：(Cl)氯仿反应、(Br)溴仿反应、(I)碘仿反应[也称为碘仿测试(iodoform test)或里本碘仿测试(Lieben iodoform test)(用于指示甲基酮的存在)][45]。

❷ 该反应大约于 1822—1870 年被首次报道[45]。

46 赫克交叉偶联
(Heck cross coupling)

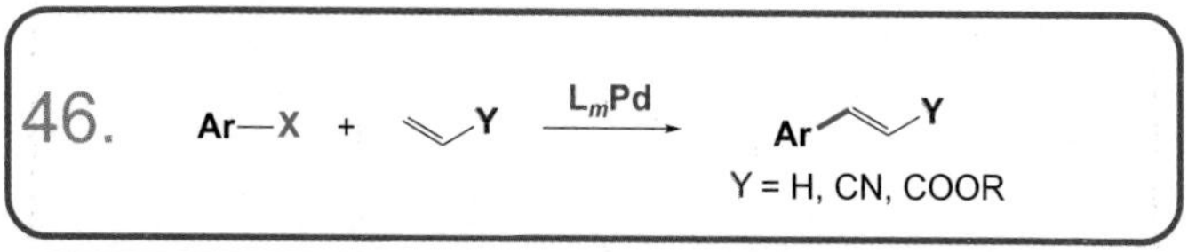

46. $Ar-X$ + $\diagup Y$ $\xrightarrow{L_mPd}$ $Ar\diagup\diagup Y$

$Y = H, CN, COOR$

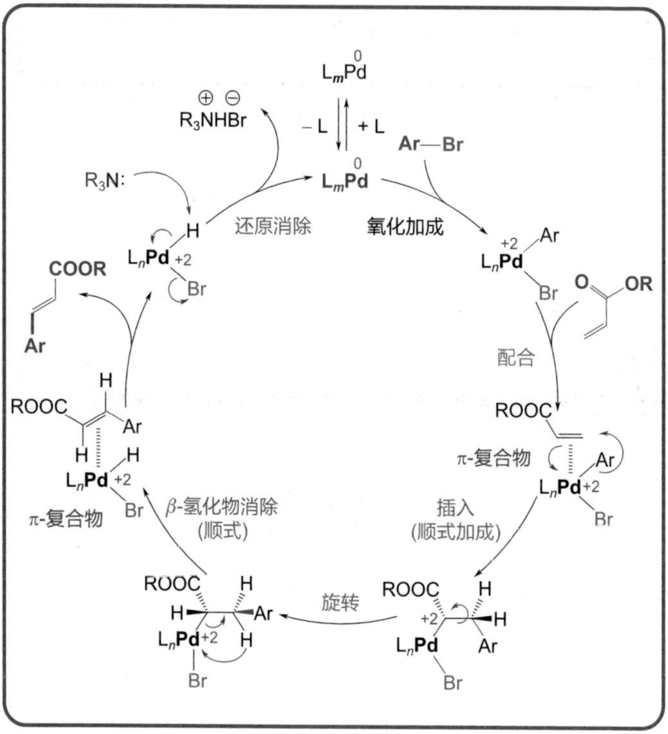

图 46.1 　赫克交叉偶联的机理 ❶

---

❶ 赫克交叉偶联，又称为赫克反应（Heck reaction）或沟吕木 - 赫克反应（Mizoroki-Heck reaction）。它是最重要的钯催化的交叉偶联反应之一，具体是指卤代烃与活化不饱和烃在钯催化下，生成偶联产物的反应（芳基卤化物和烯烃形成 C—C 键）。图中显示了简化的一般机理。

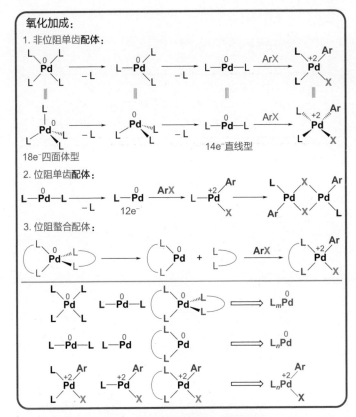

图 46.2　氧化加成步骤的一般说明 ❶

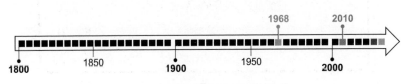

图 46.3　赫克交叉偶联反应的发现 ❷

---

❶ 氧化加成步骤在文献中以几种方式表示。按照催化剂特征分类：1. 无(少)位阻单齿配体；2. 大位阻单齿配体；3. 位阻螯合(双齿)配体。L_mPd 或 L_nPd 将作为一般情况下的简写[2a]。

❷ 该反应大约于 1968 年被首次报道[46a, 46b]。2010 年，鉴于在 Pd 催化交叉偶联反应的研究成就，Richard F. Heck、Ei-ichi Negishi 和 Akira Suzuki 共同获得了诺贝尔化学奖[46c]。

47 黑尔 – 福尔哈德 – 泽林斯基反应
(Hell-Volhard-Zelinsky reaction)

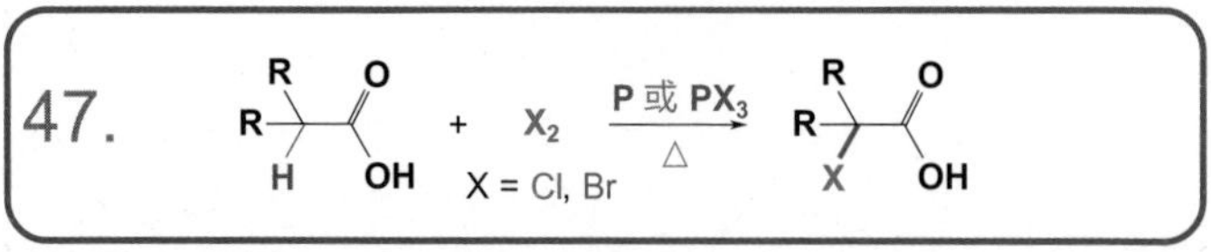

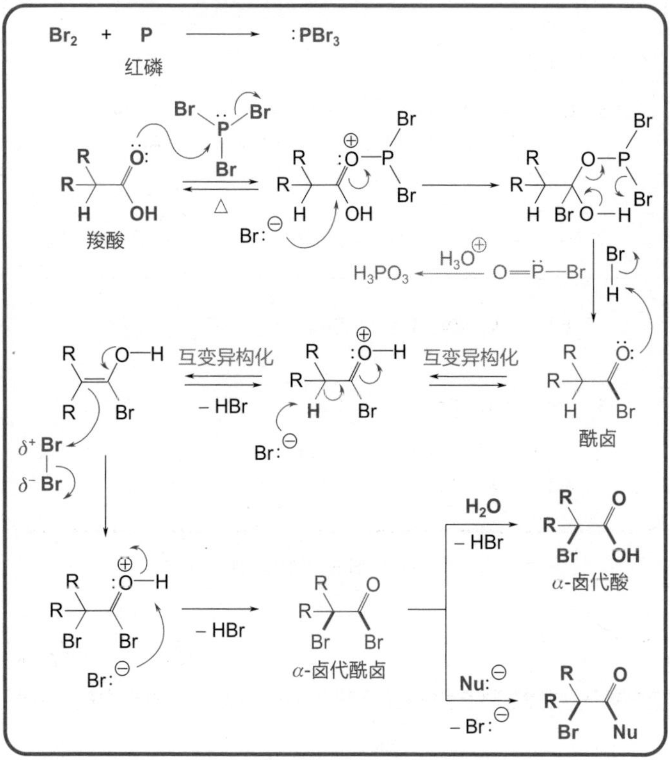

图 47.1 黑尔 – 福尔哈德 – 泽林斯基反应的机理 ❶

---

❶ 黑尔 - 福尔哈德 - 泽林斯基反应，又称为黑尔 - 福尔哈德 - 泽林斯基卤化（Hell-Volhard-Zelinsky halogenation）[译者注：是指羧酸与卤素在催化量的三溴化磷、三氯化磷（或磷和卤素）或碘等试剂的作用下，α- 氢被卤素取代生成 α- 卤代羧酸的反应。控制卤素用量，可以得到一元或多元的卤代酸]。该反应是脂肪族亲电取代的一种（参见第 3 个反应机理），从机理上而言也与卤仿反应有关（参见第 45 个反应机理）。

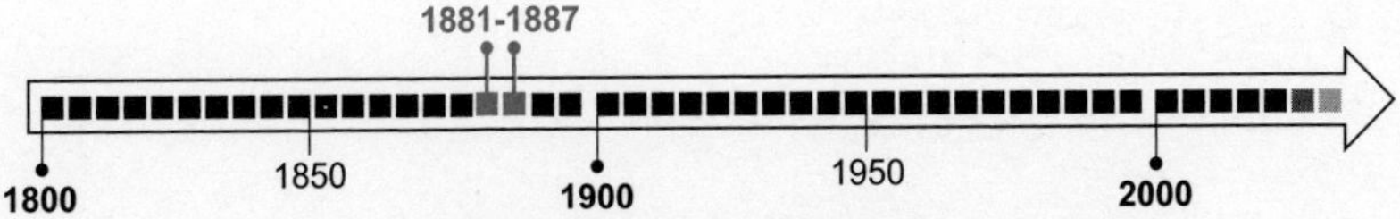

图 47.2　黑尔－福尔哈德－泽林斯基反应的发现 ❶

---

❶ 黑尔-福尔哈德-泽林斯基反应最早于 1881 年由 Hell[47a]提出，而后在 1887 年左右分别由 Volhard[47b]和 Zelinsky[47c]提出。

48 桧山交叉偶联
(Hiyama cross coupling)

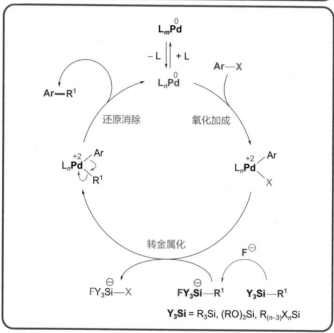

图 48.1 桧山交叉偶联的机理 ❶

---

❶ 桧山交叉偶联是一种钯催化的交叉偶联反应，使芳基卤化物和有机硅烷形成 C—C 键。为了便于理解，图中显示了简化的一般机理。

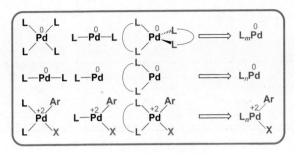

图 48.2　氧化加成步骤 ❶

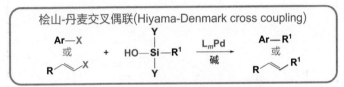

图 48.3　桧山交叉偶联的相关反应 ❷

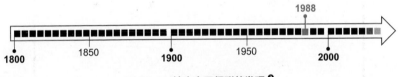

图 48.4　桧山交叉偶联的发现 ❸

---

❶ 如第 46 个反应机理所述，氧化加成步骤的表示形式可以有所不同。为了统一简化，使用了催化剂-配体配合物的简写：L_mPd 或 L_nPd[2a]。

❷ 桧山交叉偶联的一种改进称为桧山-丹麦交叉偶联[48a]。这也是一种钯催化的交叉偶联反应(使芳基卤化物和有机硅烷形成 C—C 键)。

❸ 该反应大约于 1988 年被首次报道[48b]。

49 霍夫曼消除
(Hofmann elimination)

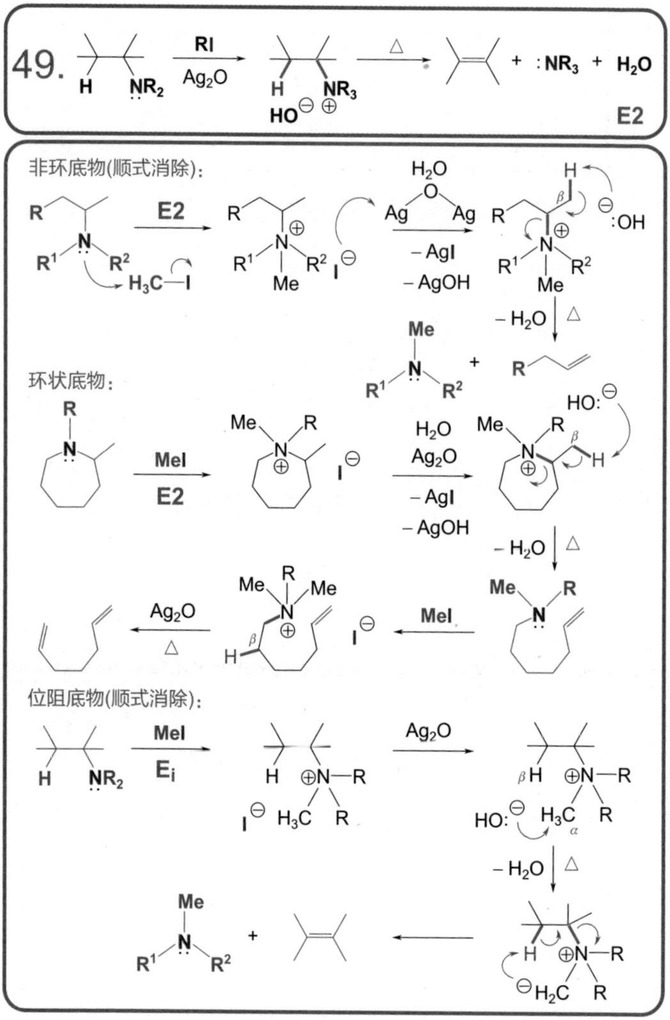

图 49.1　霍夫曼消除的机理 ❶

---

❶ 霍夫曼消除，又称为霍夫曼降解（Hofmann degradation）[译者注：是指胺与过量碘甲烷、氧化银和水共热时生成三级胺和烯烃的反应，反应中间产物为四级铵碱]。霍夫曼消除属于第 6 个反应机理所述的 β- 消除反应，与霍夫曼重排（Hofmann rearrangement）有所区别（参见第 31 个反应机理）。

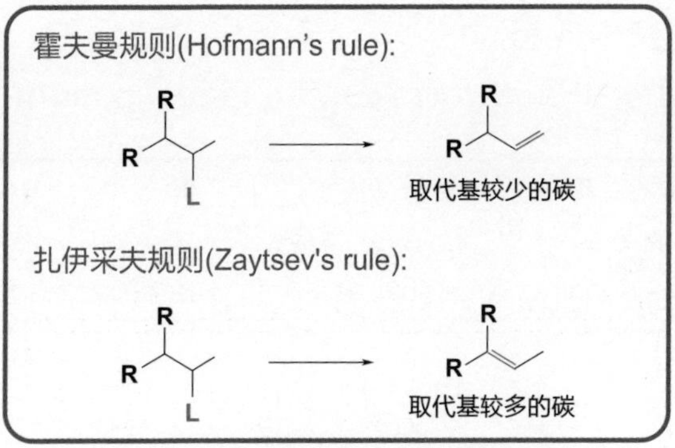

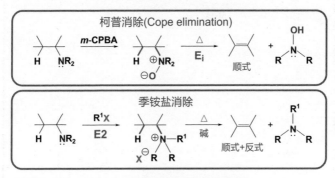

图49.2 霍夫曼规则和扎伊采夫规则 ❶

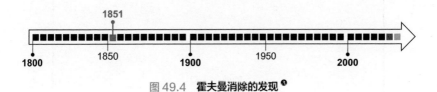

图49.3 霍夫曼消除的相关反应 ❷

图49.4 霍夫曼消除的发现 ❸

---

❶ 霍夫曼消除的产物遵循霍夫曼规则(Hofmann's rule):双键生成在取代基较少的碳上。如果双键生成在取代基较多的碳上,则符合扎伊采夫规则(Zaytsev's rule, 也称为 Saytzeff's rule)[49a]。

❷ 与霍夫曼消除有关的反应:柯普消除(E_i消除,参见第27个反应机理)、季铵盐裂解(E_2机理)等[1, 49b]。

❸ 该反应大约于1851年被首次报道[49c, 49d]。

50 霍纳-沃兹沃斯-埃蒙斯烯烃化
(Horner-Wadsworth-Emmons olefination)

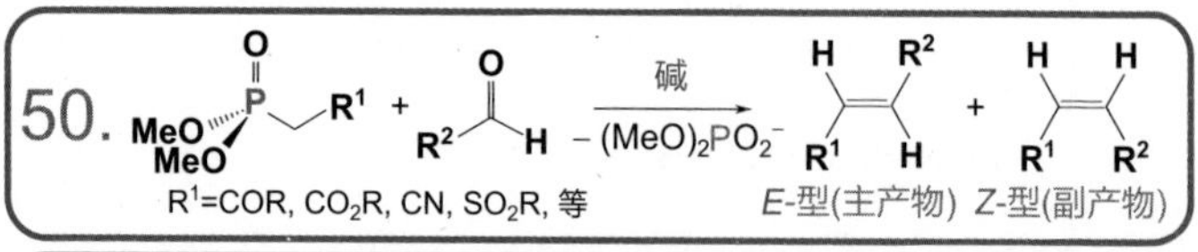

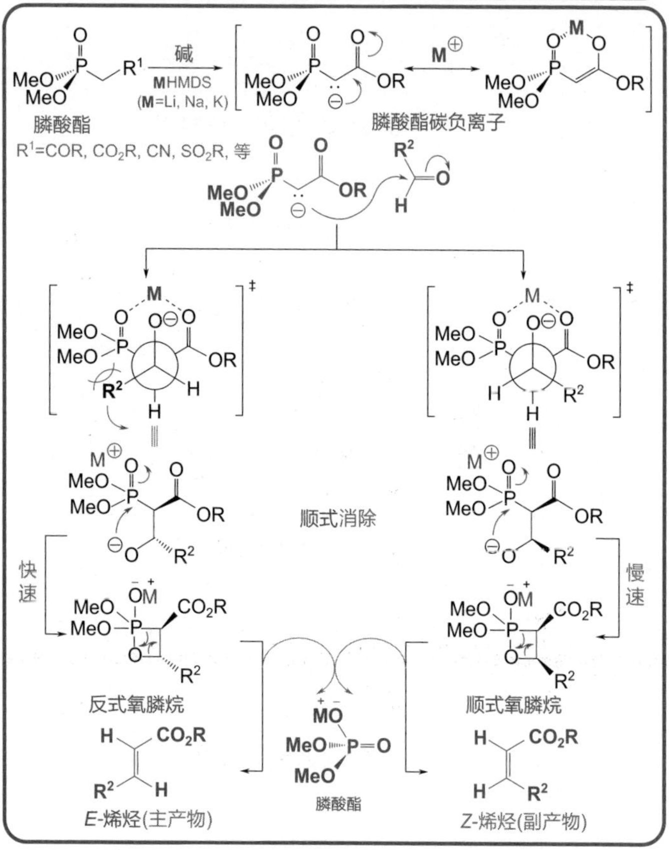

图 50.1 霍纳-沃兹沃斯-埃蒙斯烯烃化的机理 ❶

---

❶ 霍纳-沃兹沃思-埃蒙斯烯烃化,简称 HWE 反应 [译者注:是指氧化膦稳定的碳负离子与醛加成,生成 β-羟基氧化膦,而后与碱作用消除生成烯烃的反应,是维蒂希反应的改进反应]。该反应需要用到膦酸酯,膦酸酯可通过阿尔布佐夫反应 (Arbuzov reaction) 制备 (参见第 9 个反应机理)。

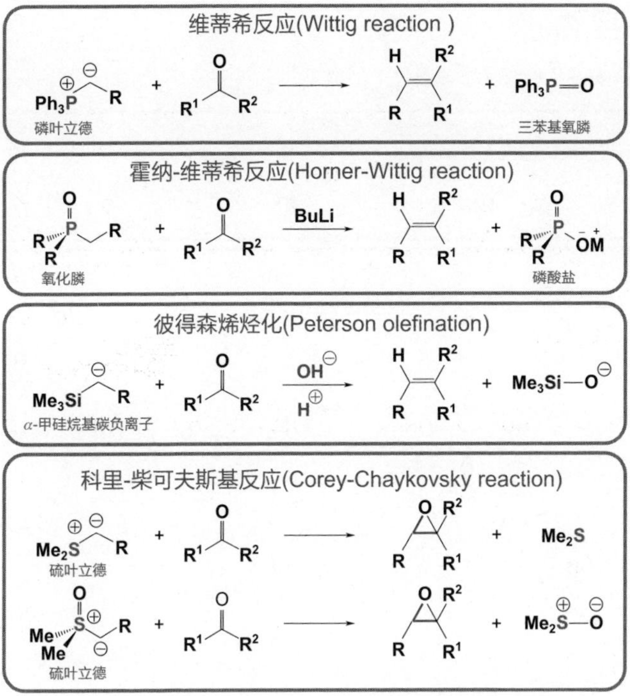

图 50.2 霍纳－沃兹沃斯－埃蒙斯烯烃化的相关反应 ❶

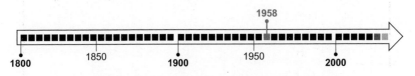

图 50.3 霍纳－沃兹沃斯－埃蒙斯烯烃化的发现 ❷

---

❶ 霍纳-沃兹沃思-埃蒙斯烯烃化反应的相关反应：维蒂希反应、霍纳-维蒂希反应[1, 50a]、彼得森烯烃化（由有机硅烷作为底物）[50b]、科里-柴可夫斯基反应（由硫叶立德作为底物，参见第 32 个反应机理）。

❷ 该反应大约于 1958 年被首次报道[50c, 50d, 50e]。

51 琼斯氧化
(Jones oxidation)

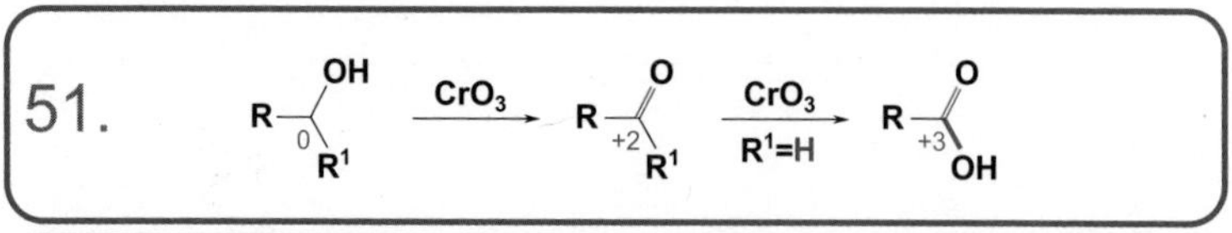

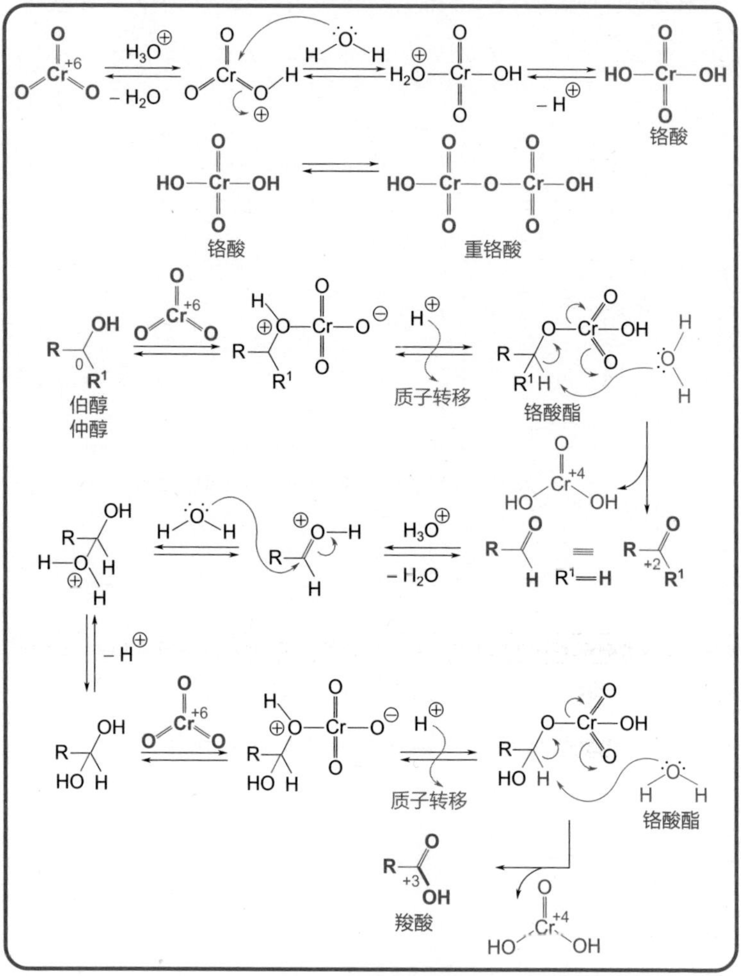

图 51.1 **琼斯氧化的机理** ❶

---

❶ 琼斯氧化是基于使用相同名称的试剂：琼斯试剂[51a]。

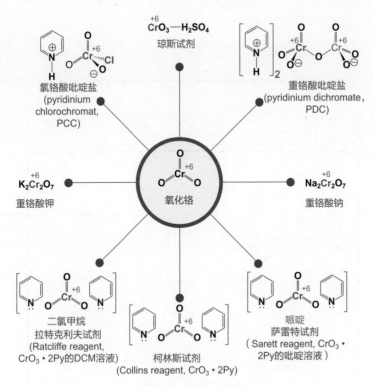

图 51.2　由氧化铬生成的各种氧化剂（Ⅵ）❶

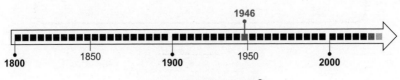

图 51.3　琼斯氧化的发现 ❷

---

❶ 由氧化铬制备铬氧化剂（Ⅵ）的实例有很多，而氯铬酸吡啶盐（PCC）[51b, 51c]是其中最重要的一种。

❷ 该反应大约于1946年被首次报道[51d]。

52 库切罗夫反应
(Kucherov reaction)

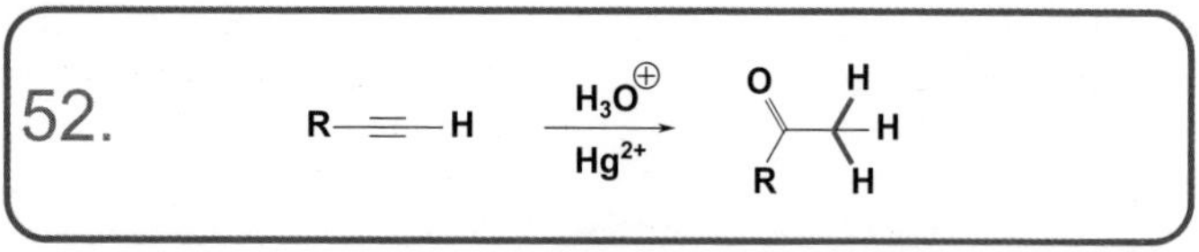

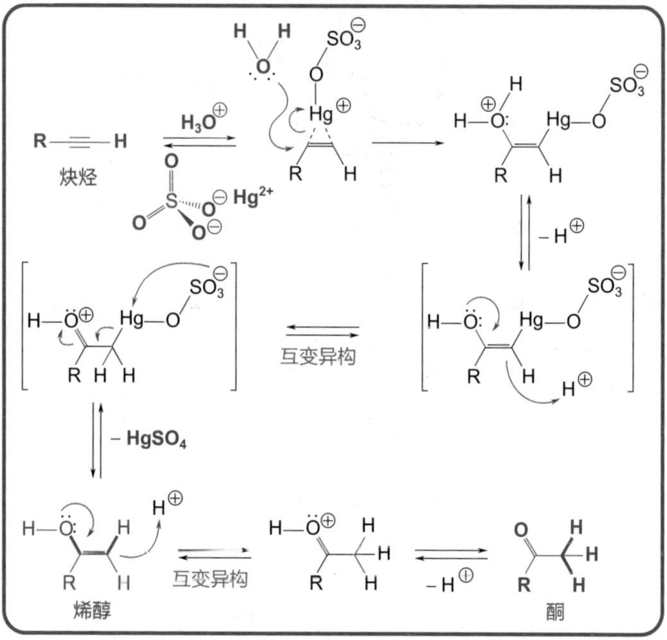

图 52.1 库切罗夫反应的机理 ❶

---

❶ 库切罗夫反应很少用其名字来称呼。[译者注：库切罗夫反应是指乙炔衍生物在含汞催化剂作用下与水加成形成羰基化合物的反应。]从反应机理上而言，属于第 1 个反应机理中亲电加成（炔烃）反应的一个实例。该反应遵循马尔可夫尼科夫规则（Markovnikov's rule，简称马氏规则）：氢（H^+ 或分子的任何其他亲电部分）加成至取代最少的碳上（或氢加成至含有更多氢的碳上）。[52a]

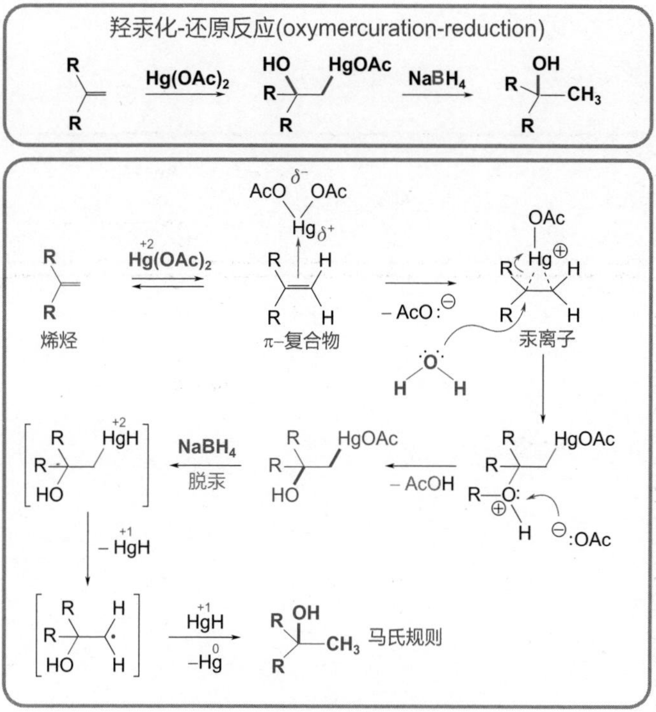

52 库切罗夫反应

图 52.2　羟汞化－还原反应的机理 ❶

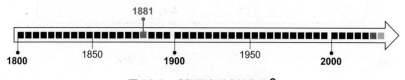

图 52.3　库切罗夫反应的发现 ❷

---

❶ 羟汞化-还原反应与库切罗夫反应相关。该反应也是主要生成醇产物的亲电加成反应，同样遵循马氏规则。需注意，硼氢化-氧化反应（参见第 20 个反应机理）会产生反马氏规则（*anti*-Markovnikov's rule）的产物：氢加成至取代最多的碳上（或氢加成至含氢最少的碳上）。

❷ 该反应大约于 1881 年被首次报道[52b]。

53 熊田交叉偶联
(Kumada cross coupling)

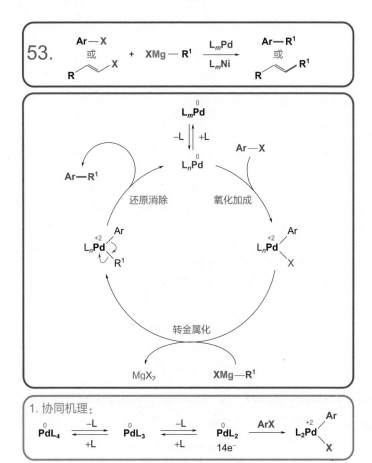

图 53.1 Pd 催化的熊田交叉偶联的机理 ❶

---

❶ 熊田交叉偶联，也称为熊田 - 玉尾交叉偶联（Kumada-Corriu cross coupling），是一种 Pd 催化的交叉偶联反应 [使用芳基卤化物和格氏试剂（有机镁化合物）构建 C—C 键]。上图中仅展示了一个简化的一般机理。注意，对低配位（14e⁻）Pd 复合物的协同氧化加成步骤较为复杂 [2a]。

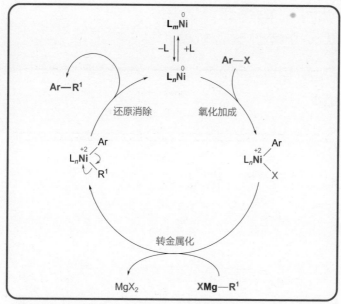

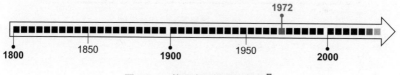

图 53.2 Ni 催化的熊田交叉偶联的机理 ❶

图 53.3 熊田交叉偶联的发现 ❷

---

❶ 熊田交叉偶联反应也可被 Ni 催化。图中显示的是对 Ni 复合物进行单电子转移（single electron transfer，SET）氧化加成反应的示例（不一定在所示示例中发挥作用）[2a]。

❷ 该反应大约于 1972 年被首次报道[53]。

54 莱伊-格里菲斯氧化
(Ley–Griffith oxidation)

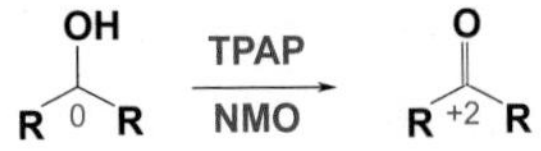

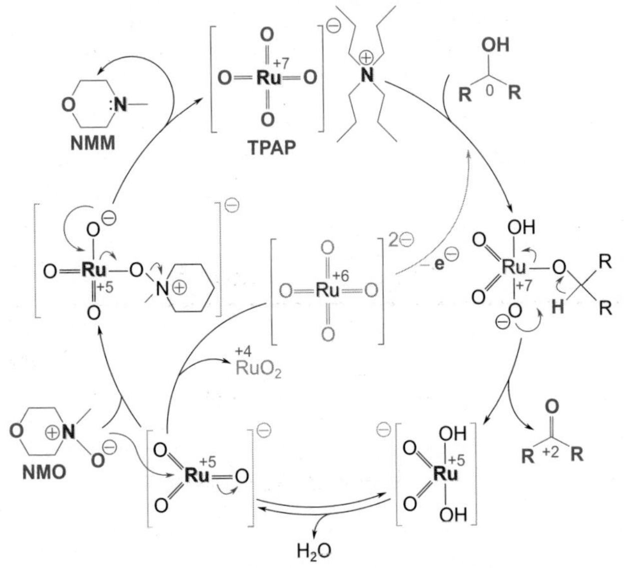

图 54.1 莱伊-格里菲斯氧化的机理 ❶

---

❶ 莱伊-格里菲斯氧化是一种基于莱伊-格里菲斯试剂过钌酸四丙基铵盐（tetrapropylammonium perruthenate，TPAP）的氧化反应，因此也被称为 TPAP 氧化[54a]。[译者注：TPAP 的氧化性弱于四氧化钌，因此底物中若含有烯键或炔键，也不会被氧化断裂，仅将伯醇氧化至醛，而不会过度氧化为羧酸。]

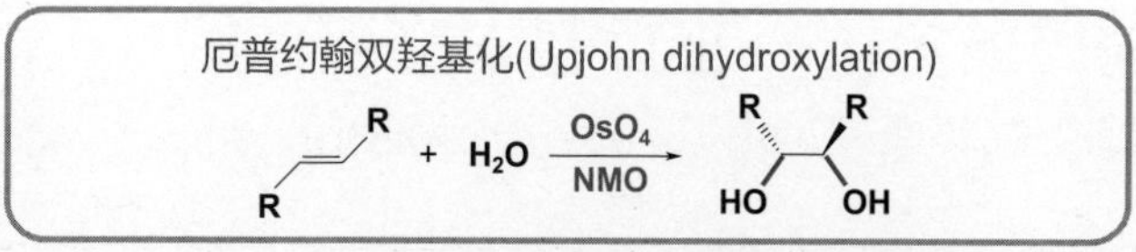

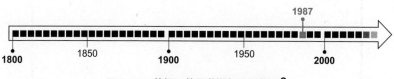

图 54.2　与莱伊－格里菲斯氧化相关的反应 ❶

图 54.3　莱伊－格里菲斯氧化的发现 ❷

---

❶ 厄普约翰双羟基化（参见第 93 个反应机理）与莱伊-格里菲斯氧化有关。
❷ 该反应大约于 1987 年被首次报道[54b]。

55 利贝斯金德-斯罗格尔交叉偶联
(Liebeskind-Srogl cross coupling)

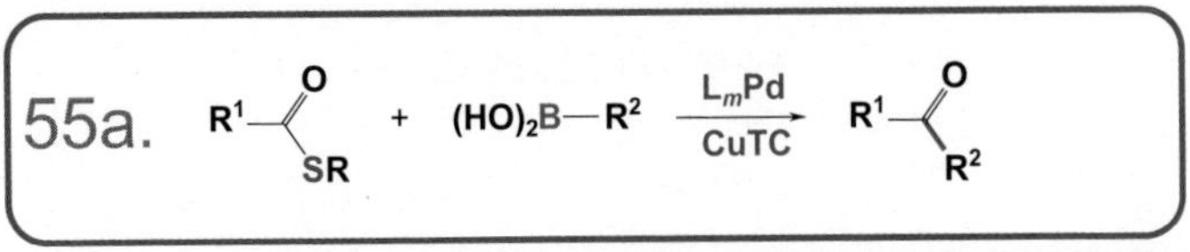

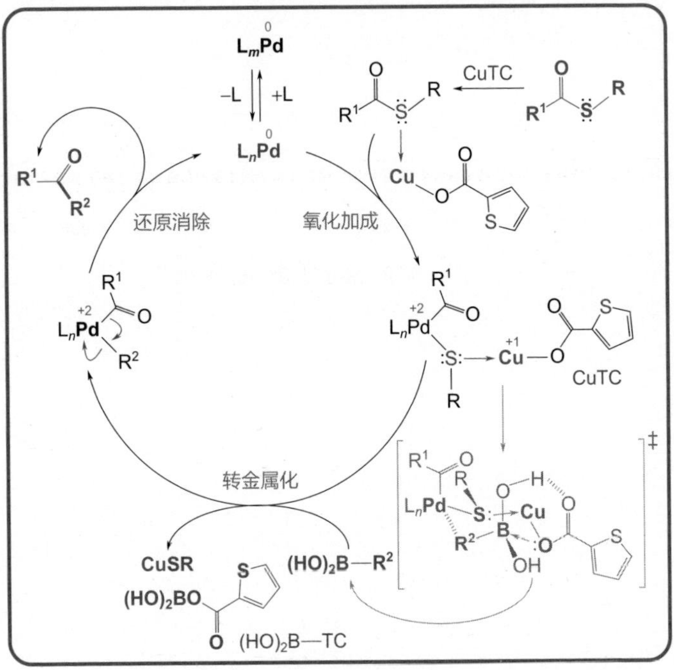

55a. $R^1\text{-C(O)-SR} + (HO)_2B-R^2 \xrightarrow{L_mPd / CuTC} R^1\text{-C(O)-}R^2$

图 55.1 **利贝斯金德-斯罗格尔交叉偶联（硫代酸酯）的机理** ❶

---

❶ 利贝斯金德-斯罗格尔交叉偶联是过渡金属 Cu 和 Pd 协同催化的硫代酸酯和硼酸间形成 C—C 键的反应。图中仅展示了简化的一般机理。

55b.

HETAr—SR
+
(HO)2B—R2
或
Bu3Sn—R2
LmPd
CuTC
HETAr—R2

LmPd
−L
+L
LnPd
HETAr—SR
HETAr—SR
CuTC
CuTC
HETAr—R2
还原消除
氧化加成
Ar
LnPd
R2
Ar
LnPd
S
CuTC
R
转金属化
CuSR
(HO)2B—TC
(HO)2B—R2

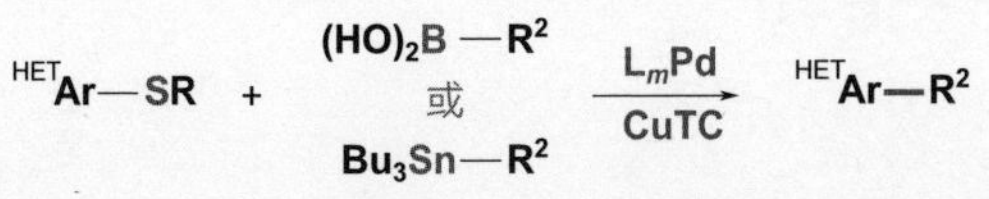

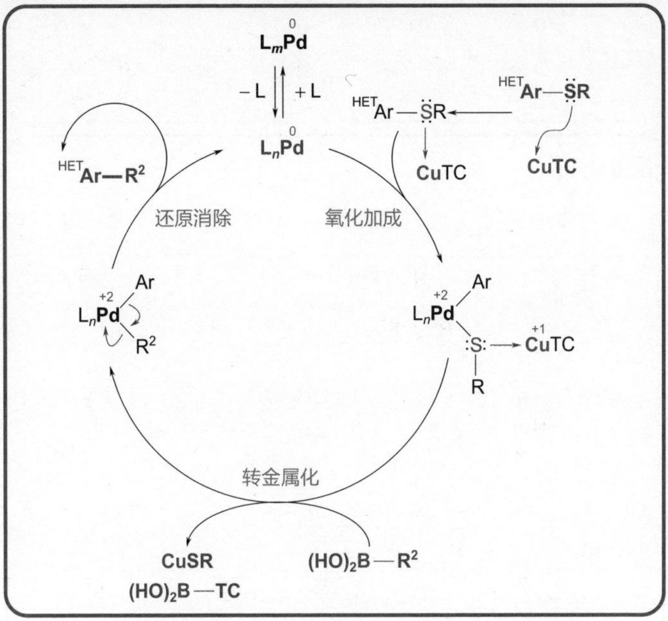

图 55.2 **利贝斯金德 - 斯罗格尔交叉偶联（硫醚）的机理** ❶

2000
1800
1850
1900
1950
2000

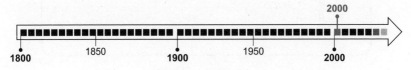

图 55.3 **利贝斯金德 - 斯罗格尔交叉偶联的发现** ❷

---

❶ 硫醚的利贝斯金德 - 斯罗格尔交叉偶联是一个特例，使用硫醚（ArSR）和硼酸或有机锡试剂构建 C—C 键（有机锡烷）。图中仅展示了简化的一般机理。

❷ 该反应大约于 2000 年被首次报道[55]。

56 曼尼希反应
(Mannich reaction)

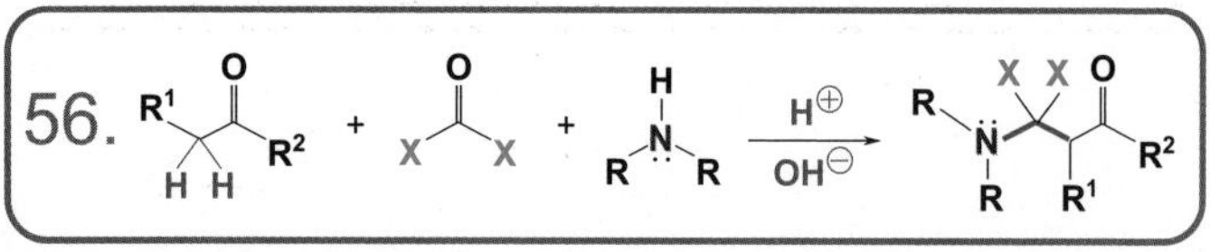

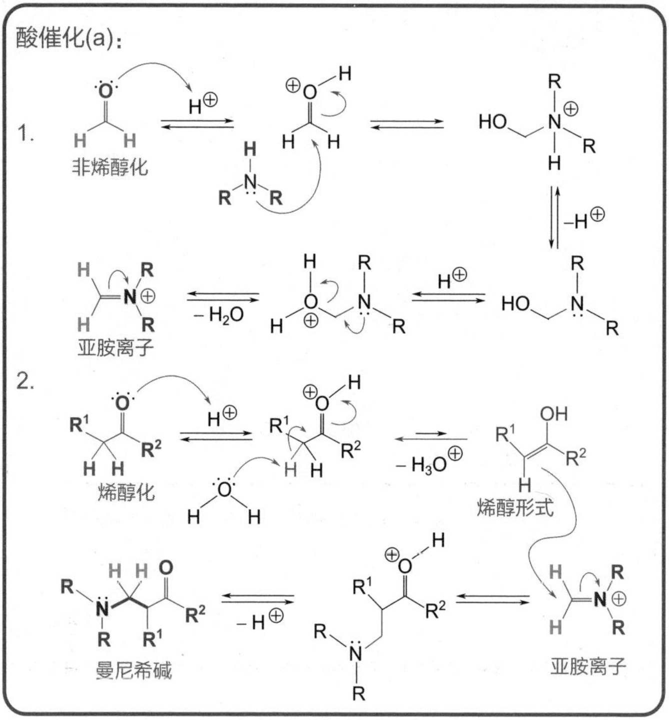

图 56.1 **曼尼希反应的机理（酸催化）**❶

---

❶ 曼尼希反应（Mannich reaction）也称为曼尼希缩合（Mannich condensation）[译者注：是指含有活泼氢的羰基化合物与甲醛和胺缩合生成 β-氨基羰基化合物的反应]。图中的三组分反应（three-component reaction，3-CR）可以在酸性介质催化下反应（形成亚胺离子中间体），终产物 β-氨基羰基也称为曼尼希碱（Mannich base）。

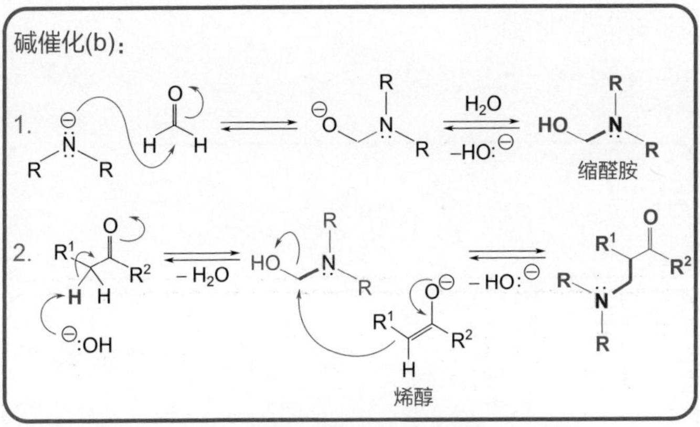

56 曼尼希反应

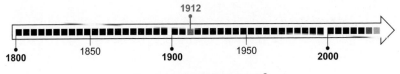

图 56.2　曼尼希反应的机理（碱催化）❶

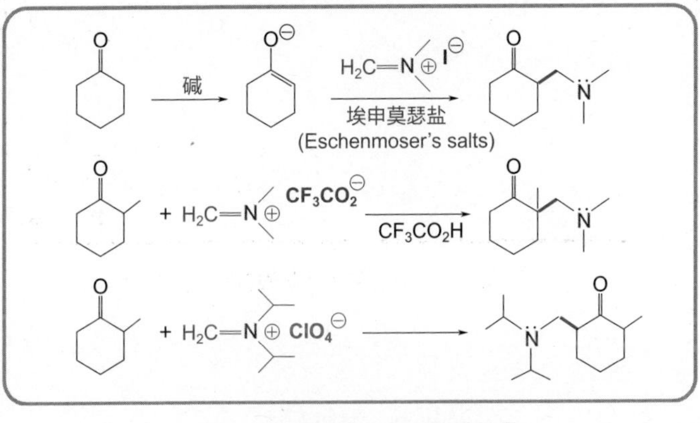

图 56.3　曼尼希反应的变体反应 ❷

图 56.4　曼尼希反应的发现 ❸

---

❶ 曼尼希反应也可在碱性介质催化下反应（形成缩醛胺中间体）。

❷ 根据预先形成的亚胺离子 [埃申莫瑟盐或伯姆盐（Böhme's salts，图中未显示）] 的可用性，曼尼希反应包括多种变体反应[56a]。

❸ 该反应大约于 1912 年被首次报道[56b]。

57 麦克默里偶联
(McMurry coupling)

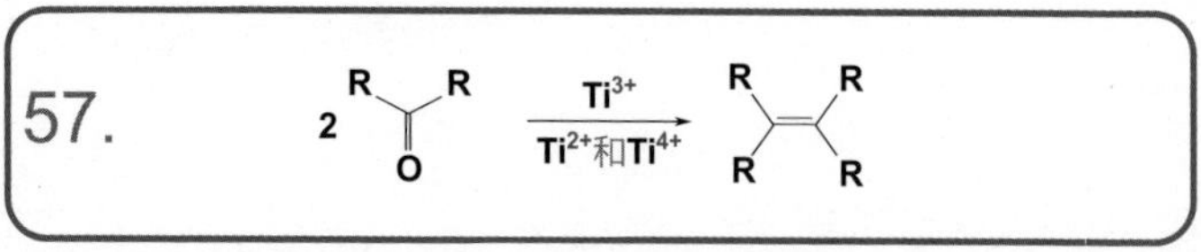

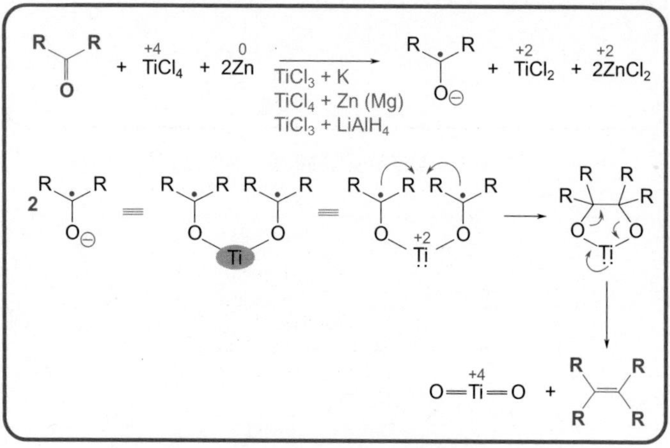

57. $2\ \underset{O}{\overset{R\ \ \ R}{C}} \xrightarrow[\text{Ti}^{2+}\text{和Ti}^{4+}]{\text{Ti}^{3+}} \underset{R\ \ \ R}{\overset{R\ \ \ R}{C=C}}$

$$\underset{O}{\overset{R\ \ R}{C}} + \overset{+4}{\text{TiCl}_4} + \overset{0}{2\text{Zn}} \xrightarrow[\substack{\text{TiCl}_3 + \text{K} \\ \text{TiCl}_4 + \text{Zn (Mg)} \\ \text{TiCl}_3 + \text{LiAlH}_4}]{} \underset{O^{\ominus}}{\overset{R\ \ R}{\dot{C}}} + \overset{+2}{\text{TiCl}_2} + \overset{+2}{2\text{ZnCl}_2}$$

图 57.1 **麦克默里偶联的机理** ❶

---

❶ 麦克默里偶联，也称为麦克默里反应（McMurry reaction），其具体机理还未被完全阐明。[译者注：麦克默里反应是指羰基化合物在低价钛（Ti）催化下制备烯烃的反应。]反应中低价钛催化剂发挥了主要作用：Ti（0）+ Ti（Ⅱ）+ Ti（Ⅲ）。

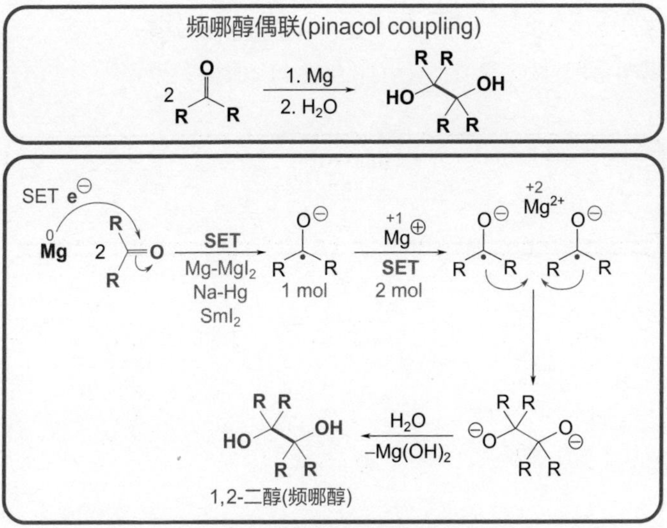

频哪醇偶联 (pinacol coupling)

$$2 \underset{R}{\underset{|}{R-C(=O)}} \xrightarrow[2.\ H_2O]{1.\ Mg} \underset{R\ R}{\underset{|\ \ |}{HO-C-C-OH}}$$

图 57.2 频哪醇偶联的机理 ❶

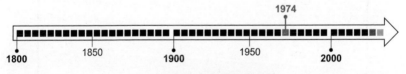

图 57.3 麦克默里偶联的发现 ❷

---

❶ 频哪醇偶联经历了单电子转移 (SET) 机理[57a, 57b]，该反应与麦克默里偶联及酮醇缩合 (acyloin condensation) 有关 (参见第 7 个反应机理)。请勿将频哪醇重排与第 76 个反应机理所述的频哪醇 - 频哪酮重排相混淆。

❷ 该反应大约于 1974 年被首次报道[57c]。

58 麦尔外因-庞多夫-维利还原
(Meerwein-Ponndorf-Verley reduction)

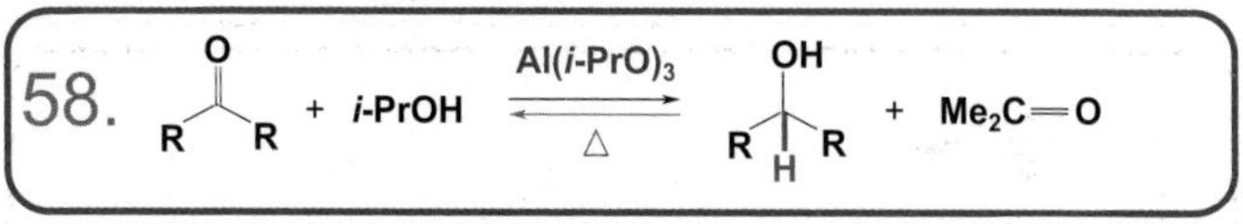

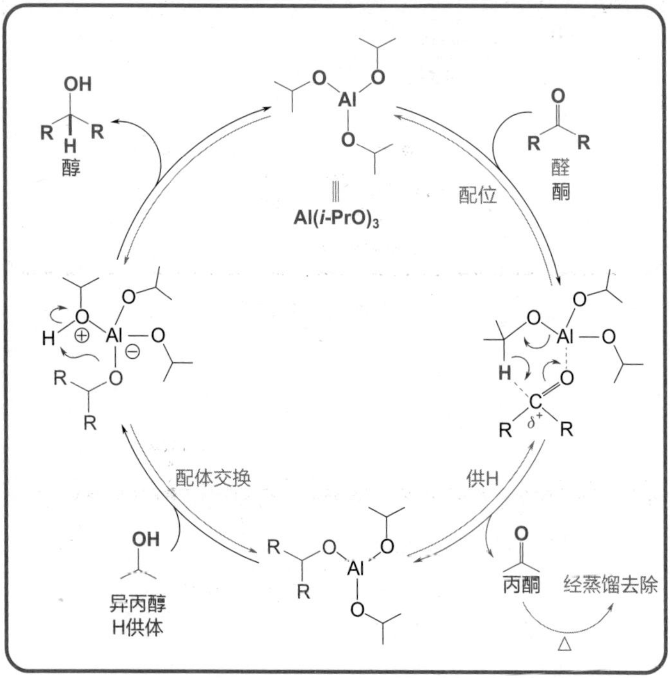

$$58. \quad \underset{R}{\overset{O}{\underset{\|}{C}}}\!\!R + \text{\textit{i}-PrOH} \xrightleftharpoons[\triangle]{\text{Al}(\textit{i}\text{-PrO})_3} \underset{R}{\overset{OH}{\underset{H}{C}}}\!\!R + \text{Me}_2\text{C}=\text{O}$$

图 58.1 麦尔外因-庞多夫-维利还原的机理 ❶

---

❶ [译者注：麦尔外因-庞多夫-维利还原（MPV reduction）是指酮在异丙醇中被 Al(*i*-PrO)$_3$ 还原为醇的反应。]该反应是可逆的，其逆反应为欧芬脑尔氧化（Oppernauer oxidation）。通过从反应混合物中除去生成的丙酮（通过蒸馏），可使平衡向还原方向移动。

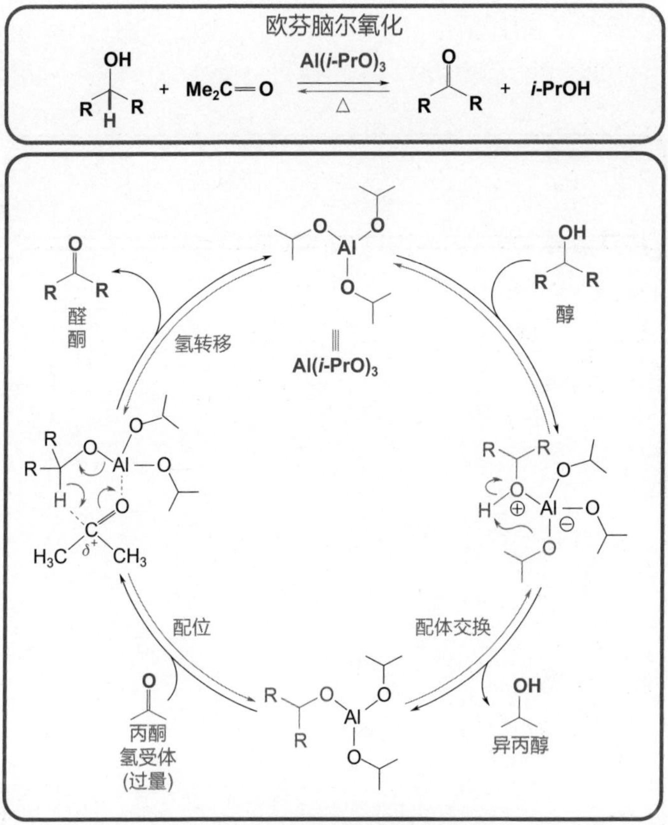

58 麦尔外因 - 庞多夫 - 维利还原

欧芬脑尔氧化

图 58.2 **欧芬脑尔氧化的机理** ❶

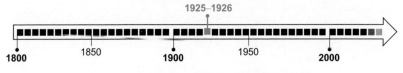

图 58.3 **麦尔外因 – 庞多夫 – 维利还原的发现** ❷

---

❶ 欧芬脑尔氧化是麦尔外因 - 庞多夫 - 维利还原的逆过程（参见第 69 个反应机理）。

❷ 麦尔外因和维利[58a, 58b]最早报道了该反应，然后庞多夫[58c]在 1926 年也对该反应进行了介绍。

59 迈克尔加成
(Michael addition)

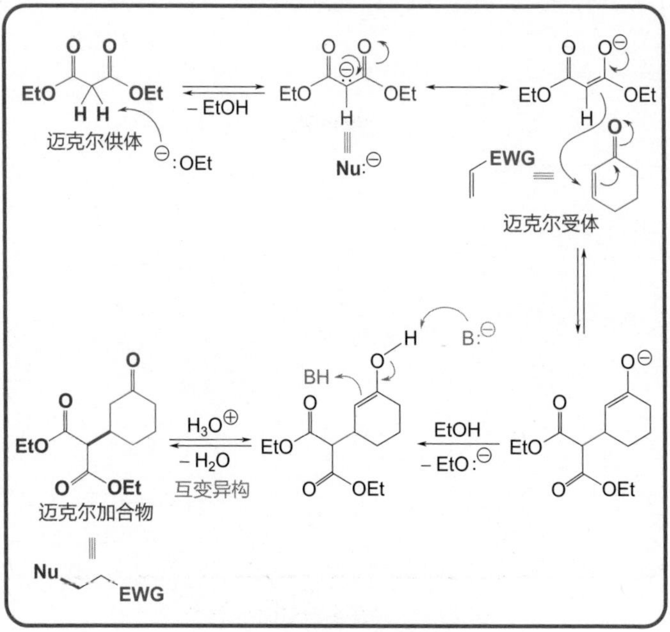

图 59.1　迈克尔加成的机理 ❶

---

❶ 迈克尔加成，也称为迈克尔共轭加成（Michael conjugate addition）或迈克尔反应（Michael reaction）[译者注：是指在碱催化下能生成亲核碳负离子的化合物与亲电的共轭体系（电子受体）所进行的共轭加成反应]。该反应是有机化学中最重要的反应之一，其反应产物被称为迈克尔加合物。

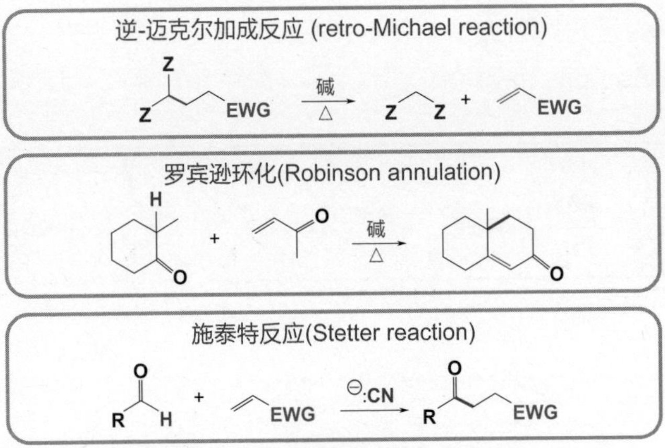

59 迈克尔加成

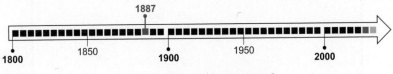

图 59.2 迈克尔加成的相关反应 ❶

图 59.3 迈克尔加成反应的发现 ❷

---

❶ 迈克尔加成反应包括很多变体反应。例如，逆-迈克尔加成和罗宾逊环化（参见第 83 个反应机理）。需注意的是，施泰特反应的机理（此处未介绍）[59a]与迈克尔加成和安息香缩合有关（参见第 15 个反应机理）。

❷ 该反应大约于 1887 年被首次报道[59b]。

60 米尼希反应
(Minisci reaction)

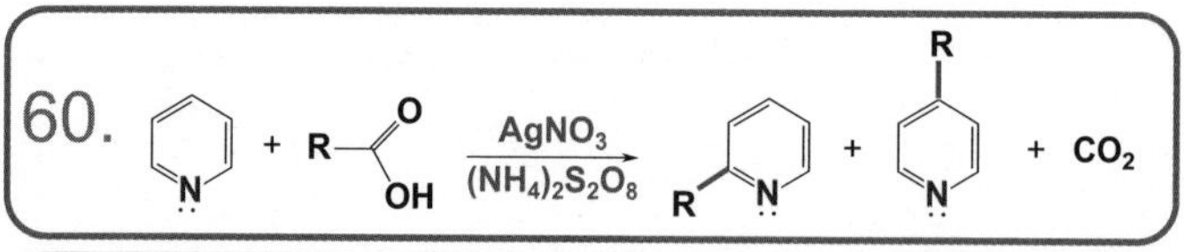

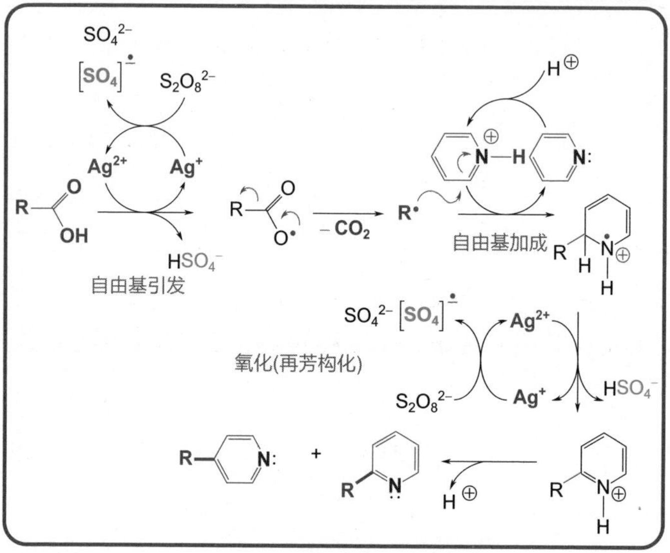

图 60.1 米尼希反应的机理 ❶

---

❶ 米尼希反应是一类自由基取代反应 [译者注：是指碳自由基加成至质子化的缺电子芳香杂环生成取代杂环化合物的反应]。与之紧密相关的机理实例包括 $S_{RN}1$ 机理（参见第 5 个反应机理）、巴顿脱羧（Barton decarboxylation）反应（参见第 12 个反应机理）和沃尔 - 齐格勒反应（Wohl-Ziegler reaction）（参见第 99 个反应机理）。

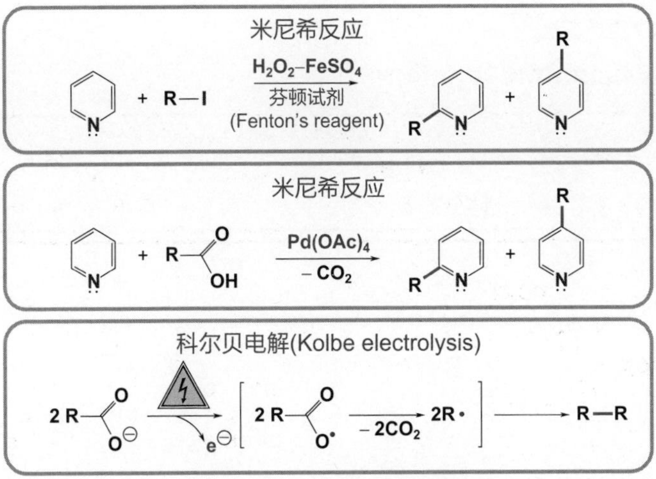

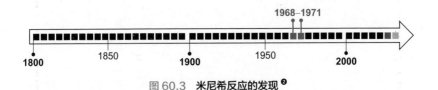

图 60.2 米尼希反应的变体反应 ❶

图 60.3 米尼希反应的发现 ❷

---

❶ 根据自由基来源的不同[如芬顿试剂[60a]和烷基碘、醋酸铅(Ⅳ)[60b]和羧酸],米尼希反应包括几种变体反应。科尔贝电解反应(也称为科尔贝反应)与该反应相关[60c]。

❷ 该反应大约于1968—1971年间被首次报道[60d, 60e]。

61 光延反应
(Mitsunobu reaction)

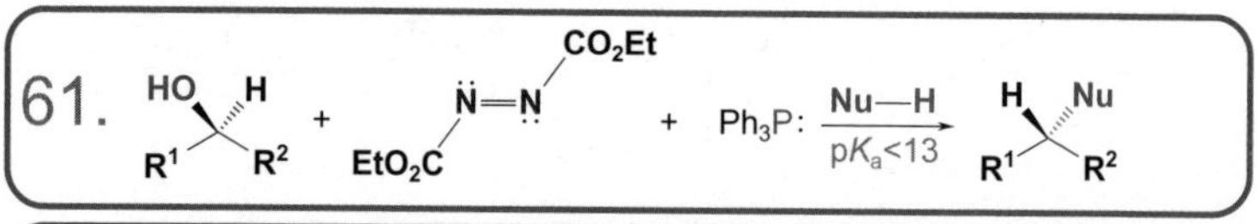

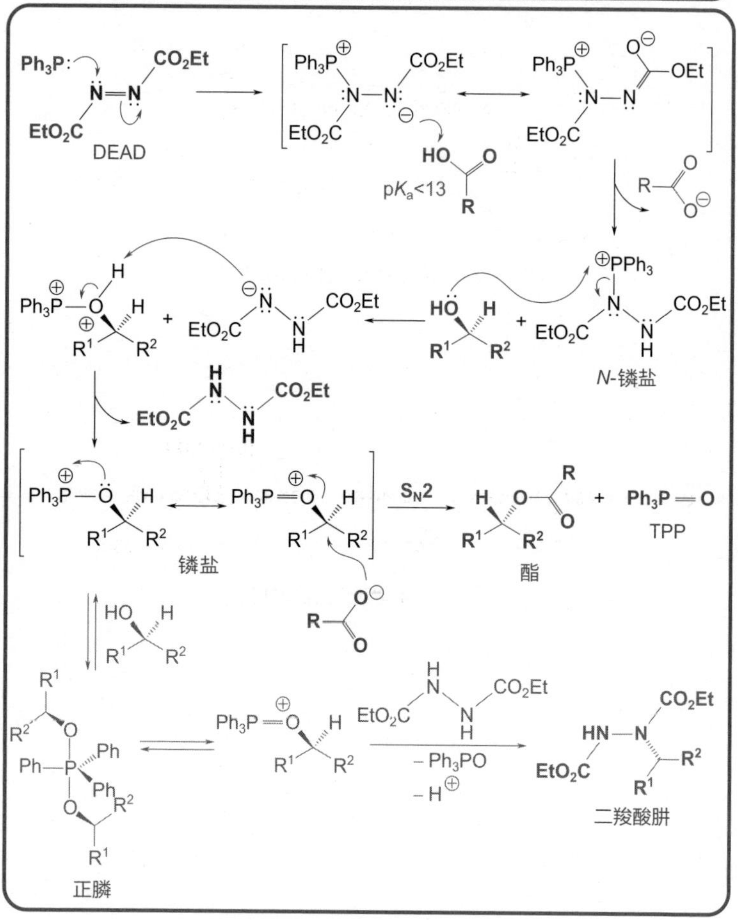

图 61.1 **光延反应的机理** ❶

---

❶ [译者注：光延反应是指伯醇或仲醇羟基在偶氮二甲酸二乙酯(DEAD)和三苯基膦的作用下被亲核试剂取代的反应，同时与羟基相连碳原子的构型发生翻转。] 光延反应的机理非常复杂，与第 2 个反应机理中所述的(脂肪族)亲核取代(S_N2)有关。注意，NuH 的 pK_a 值通常应 <13[61a]。

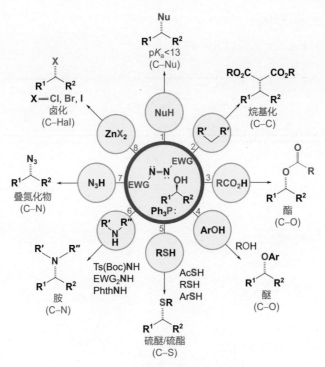

图 61.2 光延反应的合成多功能性 ❶

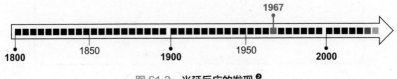

图 61.3 光延反应的发现 ❷

---

❶ 光延反应具有广泛的合成应用，可使用不同的亲核试剂（Nu）将醇转化为多种产物：1. R-Nu，pK_a<13；2. 烷基化产物；3. 酯；4. 醚；5. 硫醚或硫酯；6. 胺，C—N；7. 叠氮化物；8. 卤代烷等[61b, 61c]。

❷ 光延反应大约于 1967 年被首次报道[61d, 61e]。

62 宫浦硼化
(Miyaura borylation)

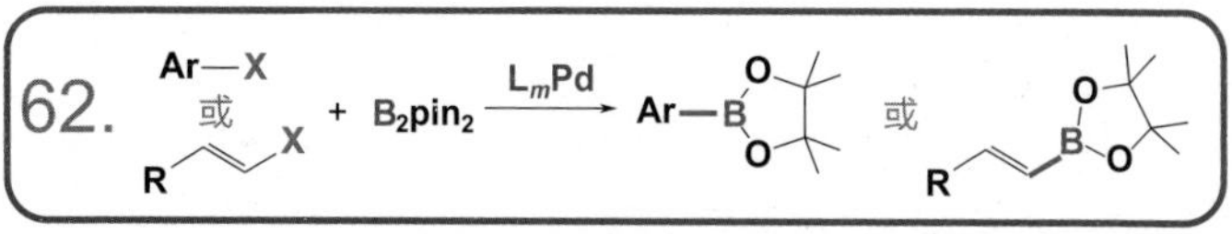

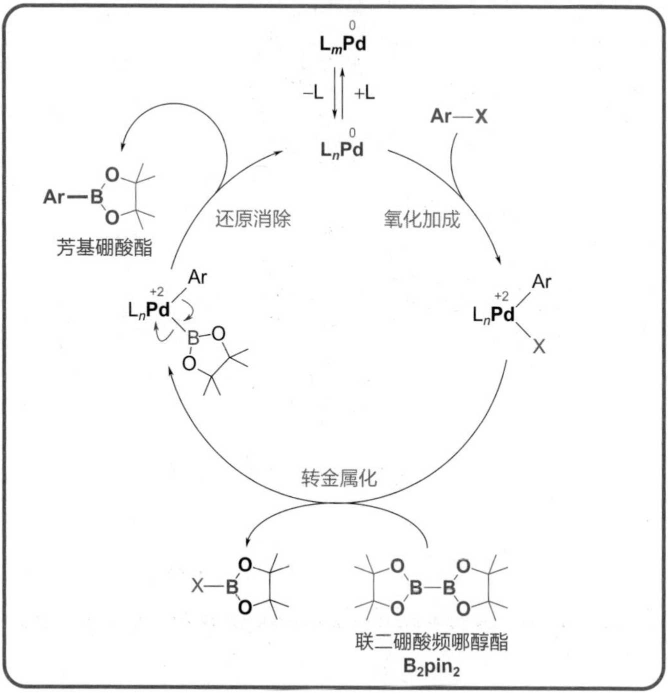

图 62.1 **宫浦硼化反应的机理** ❶

---

❶ 宫浦硼化是一种 Pd 催化的交叉偶联反应,是指芳基或烯基卤代物与联二硼酸频哪醇酯在 Pd 催化下发生偶联反应生成相应的硼酸频哪醇酯,构建 C—B 键的反应[62a]。图中仅显示了简化的一般机理。该反应合成的硼酸酯(及其相关的硼酸)是有机化学和药物化学中最重要的试剂之一。

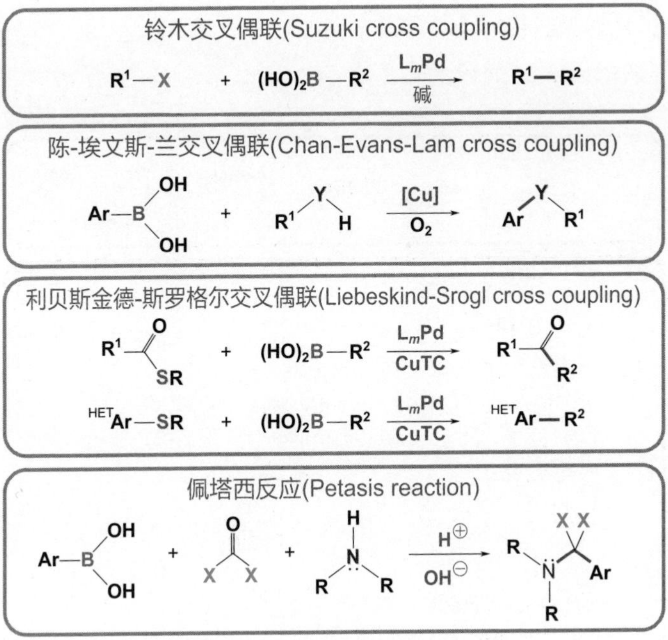

铃木交叉偶联(Suzuki cross coupling)

$$R^1-X + (HO)_2B-R^2 \xrightarrow[\text{碱}]{L_mPd} R^1-R^2$$

陈-埃文斯-兰交叉偶联(Chan-Evans-Lam cross coupling)

$$Ar-B(OH)_2 + R^1-Y-H \xrightarrow[O_2]{[Cu]} Ar-Y-R^1$$

利贝斯金德-斯罗格尔交叉偶联(Liebeskind-Srogl cross coupling)

$$R^1-C(O)-SR + (HO)_2B-R^2 \xrightarrow[\text{CuTC}]{L_mPd} R^1-C(O)-R^2$$

$$^{HET}Ar-SR + (HO)_2B-R^2 \xrightarrow[\text{CuTC}]{L_mPd} {}^{HET}Ar-R^2$$

佩塔西反应(Petasis reaction)

图 62.2 硼酸和硼酸酯的合成应用 ❶

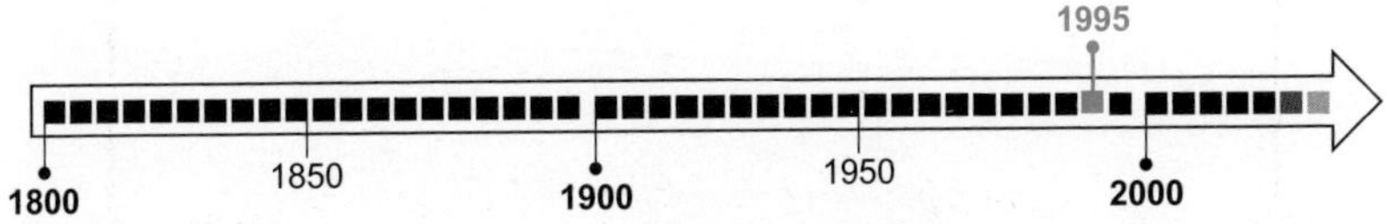

图 62.3 宫浦硼化反应的发现 ❷

---

❶ 许多关键的交叉偶联反应都使用硼酸酯(及其相关的硼酸):铃木交叉偶联(参见第89个反应机理)、陈-埃文斯-兰交叉偶联(参见第23个反应机理)、利贝斯金德-斯罗格尔交叉偶联(参见第55个反应机理)。佩塔西反应是机理上不同的三组分(3-CR)反应,但也使用硼酸[62b]。

❷ 该反应大约于1995年被首次报道[62c]。

63 向山氧化还原水合
(Mukaiyama RedOx hydration)

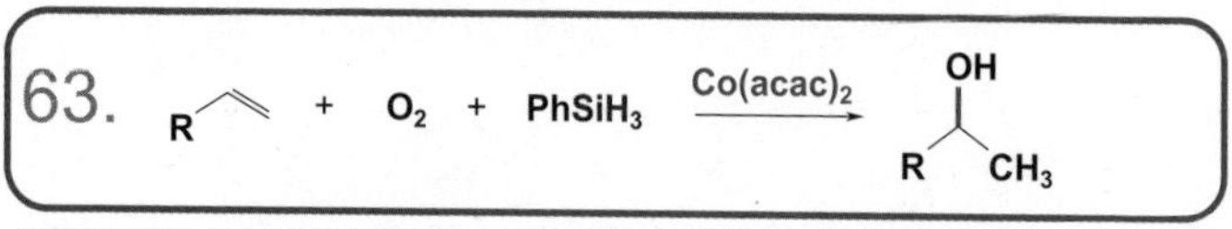

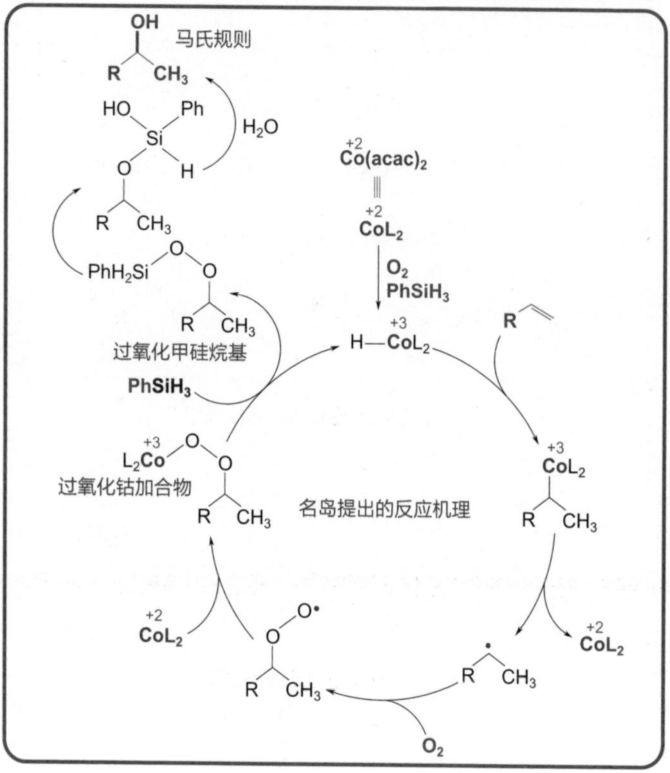

63. $R-CH=CH_2$ + O_2 + $PhSiH_3$ $\xrightarrow{Co(acac)_2}$ $R-CH(OH)-CH_3$

图 63.1 名岛提出的向山氧化还原水合的机理 ❶

---

❶ [译者注：向山氧化还原水合是指在氧气存在下，烯烃在 Co 催化下与硅烷反应生成烯烃水合物的反应。该反应遵循马氏规则。] 名岛提出了修订的向山氧化还原水合的机理[63a]。

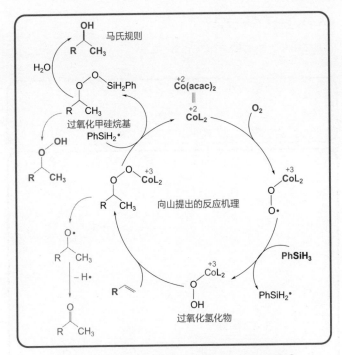

图 63.2　向山提出的向山氧化还原水合反应的机理 ❶

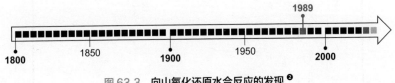

图 63.3　向山氧化还原水合反应的发现 ❷

---

❶ 向山[63b, 63c, 63d]提出了最初的向山氧化还原水合反应的机理。注意,不能将向山氧化还原水合与向山醛醇加成反应(Mukaiyama aldol addition reaction)相混淆。向山氧化还原水合是氧汞还原反应(oxymercuration-reduction reaction)的安全替代反应(参见第 20 个反应机理和第 52 个反应机理)。

❷ 该反应大约于 1989 年被首次报道[63b, 63c, 63d]。

64 纳扎罗夫环化
(Nazarov cyclization)

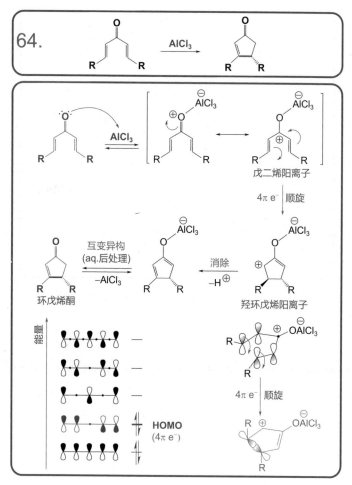

图 64.1 纳扎罗夫环化的机理 ❶

---

❶ [译者注：纳扎罗夫环化反应是指二乙烯基酮类化合物在质子酸或路易斯酸作用下重排为环戊烯酮衍生物的反应。]该反应是一种具有协调机制的周环反应，是[4π]顺旋电环化的一个实例。

	△	hv
4n	con	dis
4n+2	dis	con

图 64.2　伍德沃德 - 霍夫曼规则（Woodward-Hoffmann rules）（周环选择规则）❶

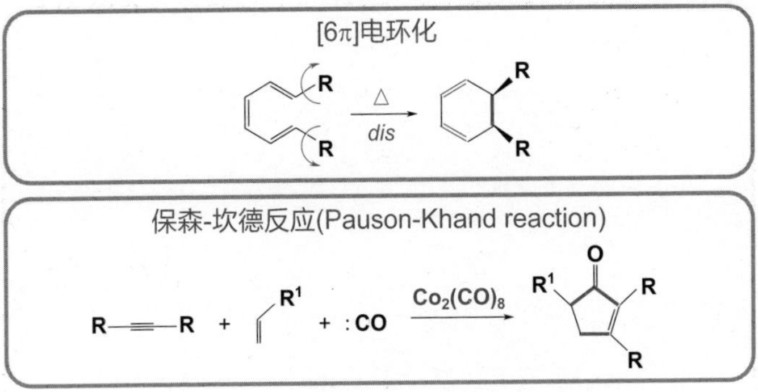

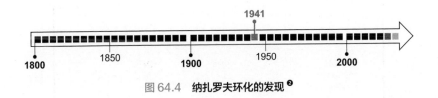

图 64.3　纳扎罗夫环化反应的相关反应❷

图 64.4　纳扎罗夫环化的发现❸

---

❶ 伍德沃德 - 霍夫曼规则（周环选择规则）[64a, 64b]适用于电环化反应。注意，纳扎罗夫环化是一个顺旋过程（$4n=4\pi$），在热状态或控制下（△）保持基态。而[6π]电环化的实例属于对旋过程（$4n+2=6\pi$），该过程保持基态（△）。在光化学条件或控制下（hv）保持激发态[64c]。

❷ 还有许多其他的[$4n$]和[$4n+2$]电环化反应的实例。保森 - 坎德反应（参见第73个反应机理）虽然经过不同的机制，但也产生了环戊烯酮。

❸ 该反应大约于1941年被首次报道[64d, 64e, 64f, 64g]。

65 尼夫反应
(Nef reaction)

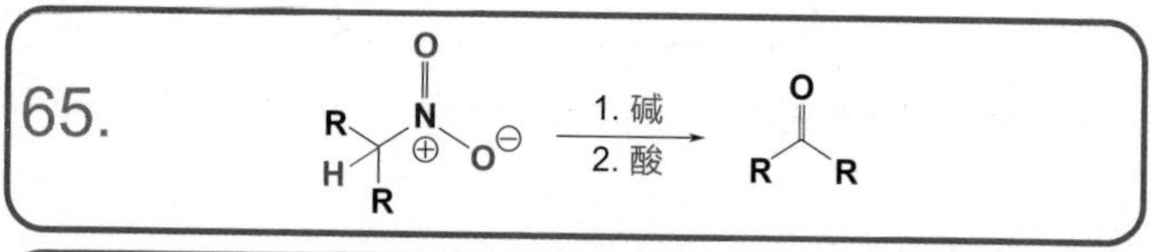

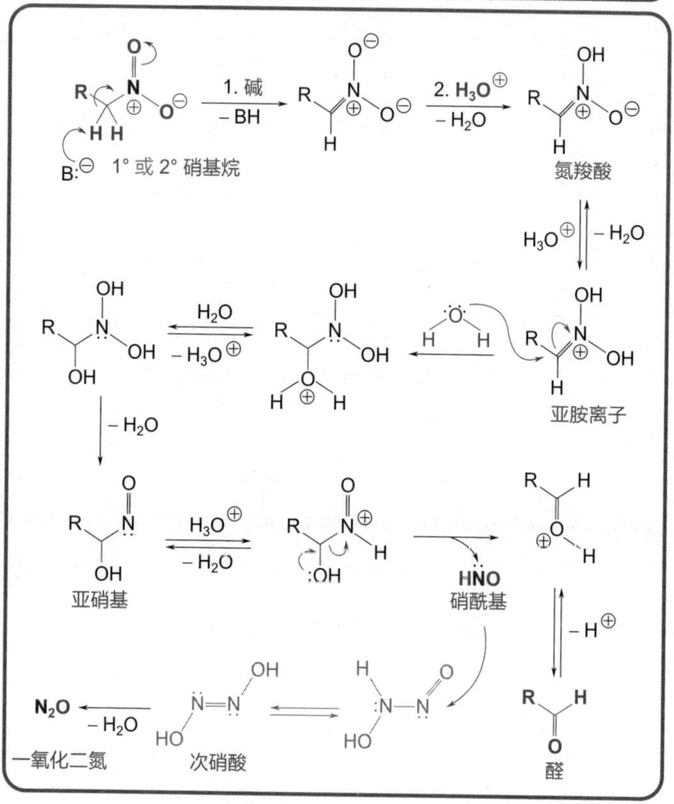

图 65.1 尼夫反应的机理（碱-酸催化）❶

---

❶ [译者注：尼夫反应是指将硝基化合物转化为羰基化合物的反应。]经典的尼夫反应是由酸催化的，生成醛和酮。需要一个碱基将1°或2°硝基烷转化为其共轭碱（氮羧酸），而3°硝基烷烃不发生该反应。

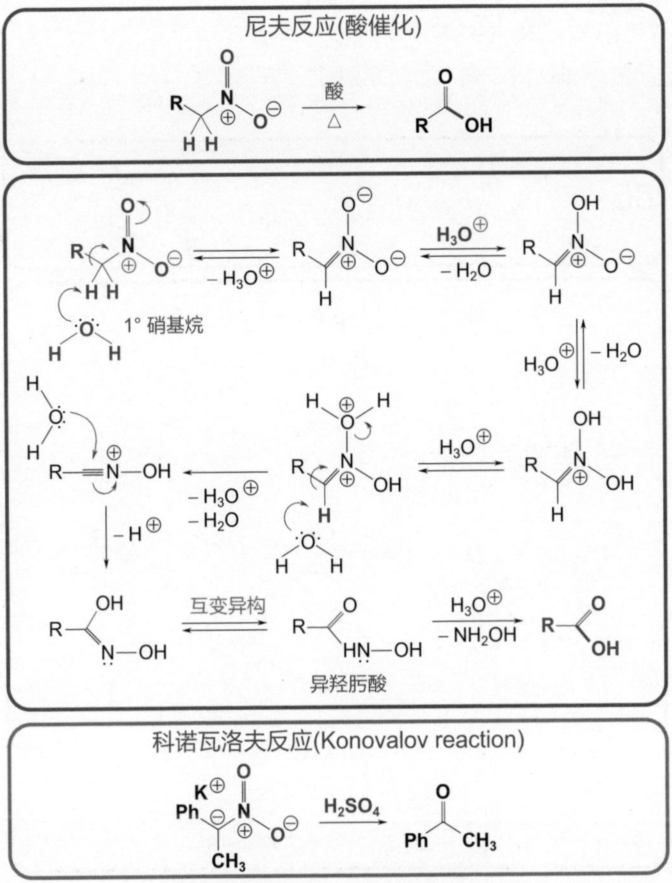

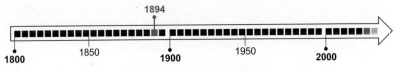

图 65.2　**尼夫反应机理（酸催化）**❶

图 65.3　**尼夫反应的发现**❷

---

❶ 如果强酸与 1° 硝基烷一起使用，则尼夫反应的机理将经历羟肟酸中间体。在这种情况下，反应会生成羧酸[1, 65a]。注意，该反应可能是由科诺瓦洛夫（Konovalov）[65b]首次报道的。

❷ 该反应大约于 1894 年被首次报道[65c, 65d]。

66 根岸交叉偶联
(Negishi cross coupling)

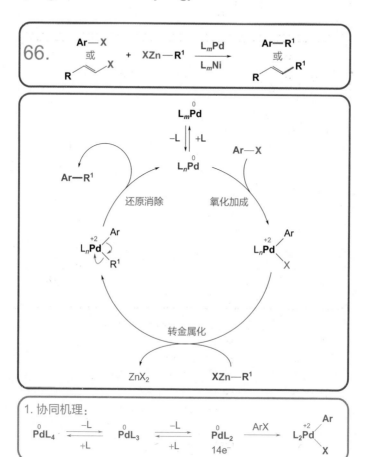

图 66.1　Pd 催化的根岸交叉偶联的机理 ❶

---

❶ 根岸交叉偶联反应是一种 Pd 或 Ni 催化的交叉偶联反应，可使芳基卤化物与有机锌化合物偶联构建 C—C 键。图中显示了一种简化的一般机理。注意，低配位（14e⁻）Pd 复合物的协同氧化加成步骤更为复杂[2a]。

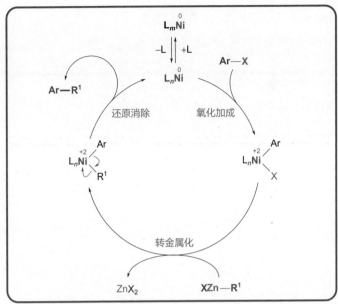

2. 单电子转移(SET)机理:

$$\overset{0}{NiL_4} \underset{+L}{\overset{-L}{\rightleftharpoons}} \overset{0}{NiL_3} \xrightarrow{ArX} [NiL_3]^{\oplus}[ArX]^{\ominus \bullet} \xrightarrow{-L} \overset{+2}{L_2Ni}\overset{Ar}{\underset{X}{\diagdown}}$$

图 66.2　Ni 催化的根岸交叉偶联的机理 ❶

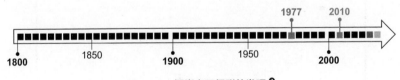

图 66.3　根岸交叉偶联的发现 ❷

---

❶ 根岸交叉偶联也可由 Ni 催化, 如经历 Ni 配合物的单电子氧化加成[2a]。

❷ 该反应大约于 1977 年被首次报道[66]。在 2010 年, Ei-ichi Negishi、Richard F. Heck 和 AkiraSuzuki 由于在 Pd 催化交叉偶联反应的成果而共同获得了诺贝尔化学奖[46c]。

67 诺里什 I 型和 II 型反应
(Norrish type I & II reaction)

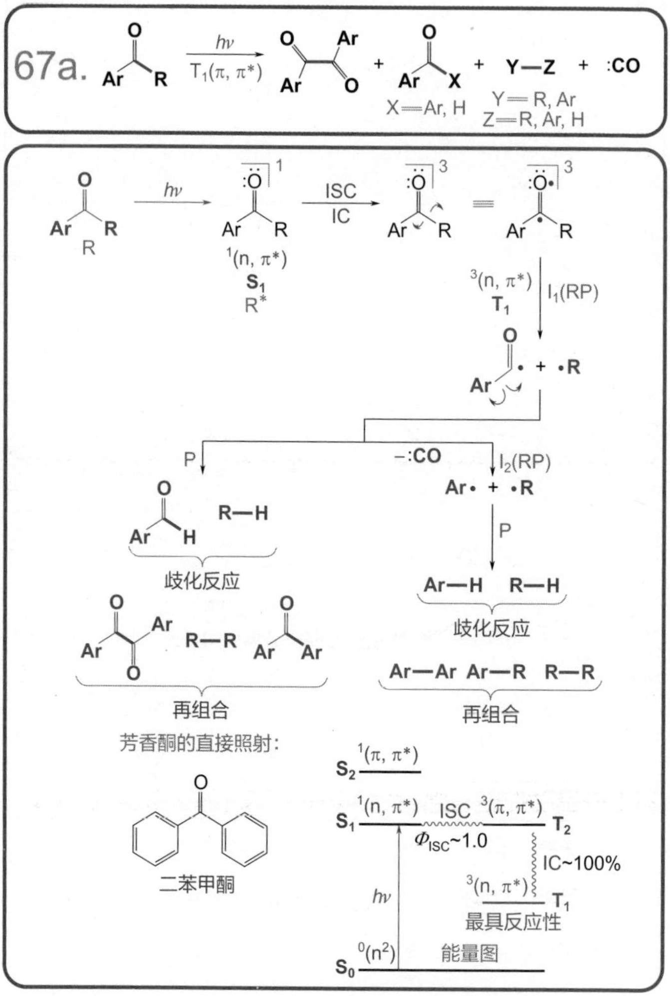

图 67.1 诺里什 I 型反应的机理 ❶

---

❶ 诺里什 I 型反应是指醛和酮的光化分解（α-裂解）反应。该反应的产物是由最初的裂解，以及随后的歧化或不同自由基的再组合而生成的。直接照射芳香酮（二苯甲酮）后，反应通常由三重激发态 $T_1 = {}^3(n, \pi^*)$ 开始[2b]。

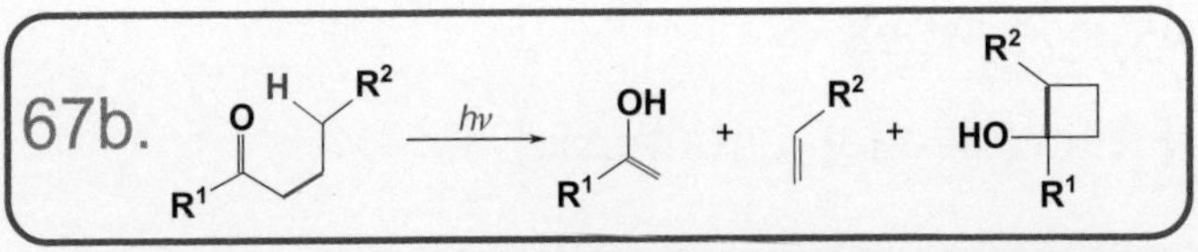

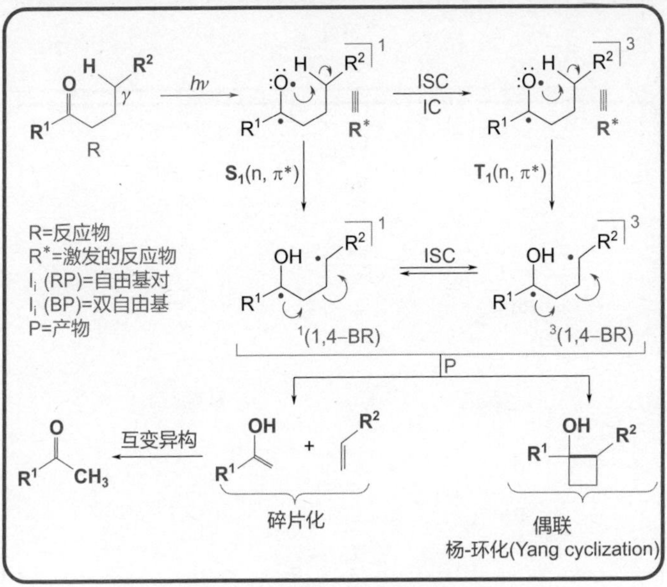

图 67.2　诺里什 II 型反应的机理 ❶

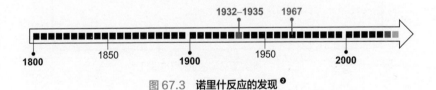

图 67.3　诺里什反应的发现 ❷

---

❶ 诺里什 II 型反应（Norrish type II reaction）是一种光化学分子内 γ-H 转移反应。[译者注：含有 γ-氢的羰基化合物的 γ-氢分子内转移至氧上形成 1,4-双自由基，随后裂解为链烯和链醇，或环化生成环丁醇。] 该反应产物可能是由 1,4-二自由基的碎裂、再组合或杨-环化形成的。反应可能发生于单重态 $S_1=^1(n, \pi^*)$ 或三重激发态 $T_1=^3(n, \pi^*)$[2b]。

❷ 诺里什 I 型和 II 型反应可能是在 1932—1935 年[67a、67b、67c、67d]或更早之前被首次报道[67e、67f]。1967 年，罗纳德·乔治·赖福德·诺里什（Ronald George Wreyford Norrish）与曼弗雷德·艾根（Manfred Eigen）和乔治·波特（George Porter）共同获得了诺贝尔化学奖[67]。

68 烯烃复分解
[oefin (alkene) metathesis]

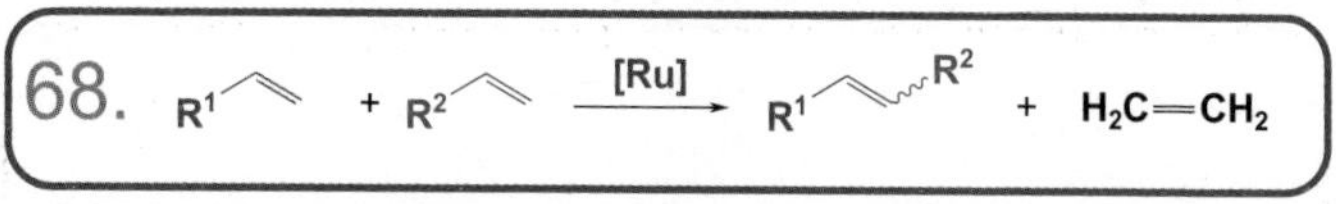

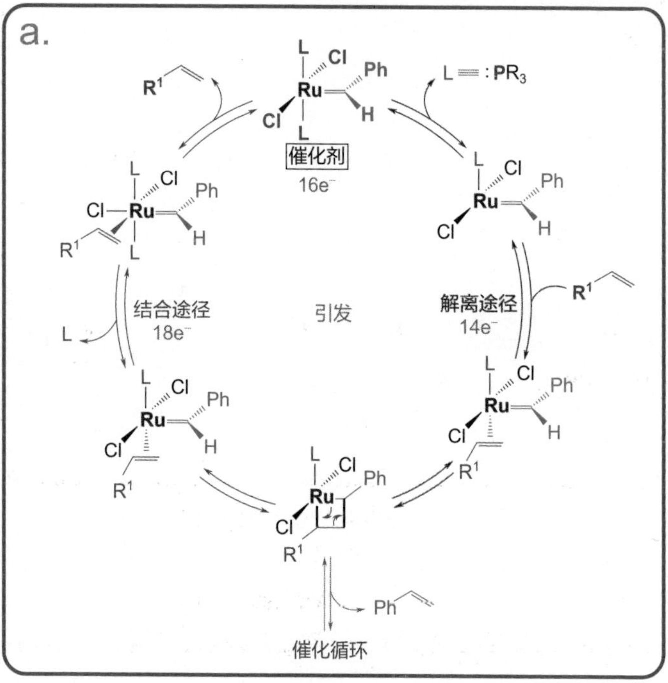

$$68. \quad R^1\text{—CH=CH}_2 + R^2\text{—CH=CH}_2 \xrightarrow{[Ru]} R^1\text{—CH=CH—}R^2 + H_2C\text{=}CH_2$$

图 68.1　烯烃复分解的机理（引发）❶

---

❶[译者注：烯烃复分解反应是指在金属催化剂的作用下，两个烯烃底物中碳碳双键连接的两个部分发生交换，生成了两个新烯烃的反应。] Ru 催化的烯烃复分解反应起始于稳定的催化剂（16e⁻）的引发循环（a）；理论上而言，其可通过解离途径（14e⁻）或结合途径（18e⁻）完成，图中未显示交换途径[68a]。

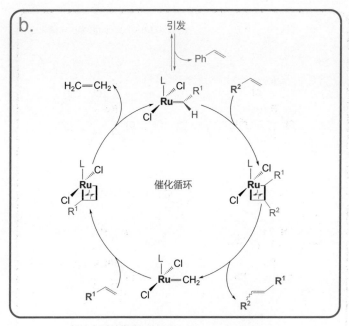

图 68.2　烯烃复分解反应的机理（催化循环）❶

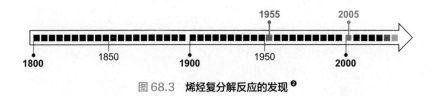

图 68.3　烯烃复分解反应的发现❷

---

❶ 苯乙烯结构部分消耗后，主要的催化循环（b）在"活性"催化剂的作用下继续进行。注意，该机理非常复杂，并且随着底物和催化剂的不同而变化很大。本图仅显示了一个简化的一般示例。

❷ 该反应大约于 1955 年被首次报道[68b, 68c]。2005 年，Yves Chauvin、Robert H. Grubbs 和 Richard R. Schrock 因在该反应领域的成果而获得了诺贝尔化学奖[68d]。

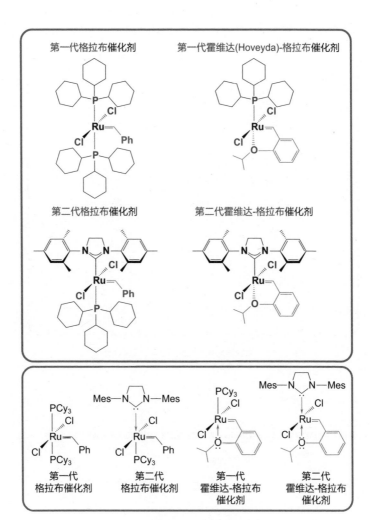

图 68.4　主要的烯烃复分解反应催化剂❶

---

❶ Ru 催化的烯烃复分解反应中最常用的催化剂是格拉布催化剂（第一代和第二代）[68e, 68f]和霍维达-格拉布催化剂（第一代和第二代）[68g]。

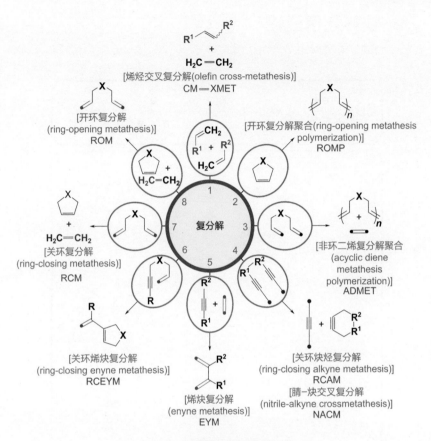

图 68.5 烯烃复分解反应的分类❶

---

❶ 烯烃复分解反应可分为：1. 烯烃交叉复分解（CM=XMET）；2. 开环复分解聚合（ROMP）；3. 非环二烯复分解聚合（ADMET）；4. 关环炔烃复分解（RCAM）和腈-炔交叉复分解（NACM）；5. 烯炔复分解（EYM）；6. 关环烯炔复分解（RCEYM）；7. 关环复分解（RCM）；8. 开环复分解（ROM）。

69 欧芬脑尔氧化
(Oppenauer oxidation)

$$69. \quad \underset{H}{\overset{OH}{\underset{|}{R-C-R}}} + Me_2C=O \xrightarrow[\Delta]{Al(i\text{-}PrO)_3} \underset{R}{\overset{O}{\underset{\|}{R-C-R}}} + i\text{-}PrOH$$

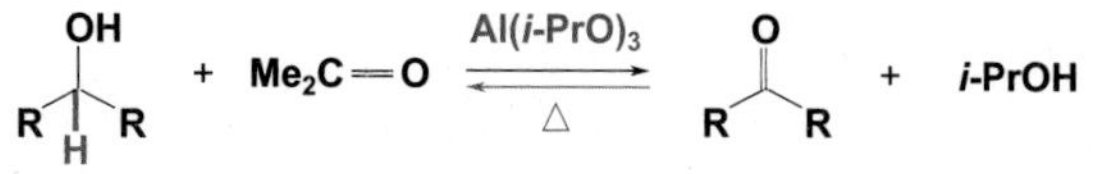

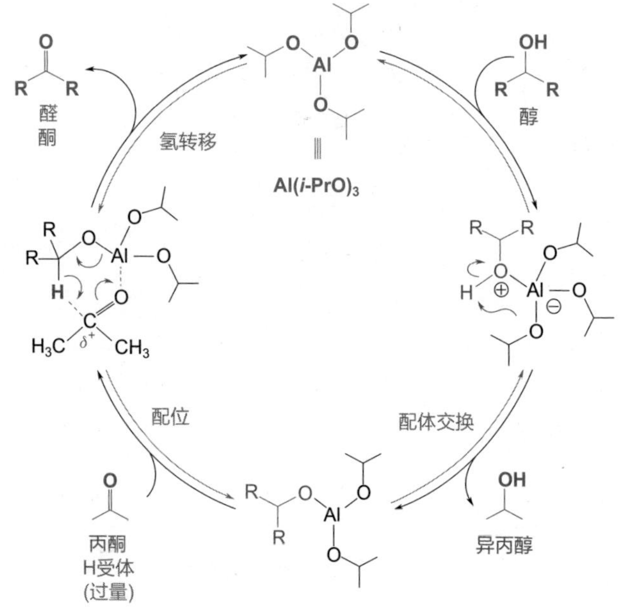

图 69.1 欧芬脑尔氧化的机理 ❶

---

❶ [译者注：欧芬脑尔氧化反应是指选择地将醇羟基氧化成羰基的反应。由于伯醇氧化后生成的醛在碱性条件下易发生羟醛缩合，因此该方法通常用于将仲醇氧化成酮。] 欧芬脑尔氧化是可逆的，其逆向还原反应被称为麦尔外因-庞多夫-维利（Meerwein-Ponndorf-Verley，MPV）还原。通过添加过量的丙酮，平衡可以向氧化方向移动。

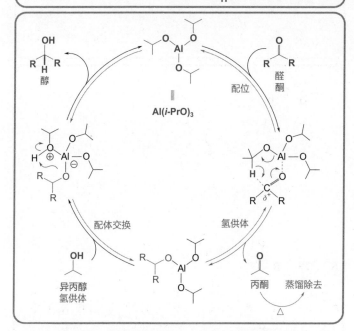

图 69.2　麦尔外因－庞多夫－维利还原的机理❶

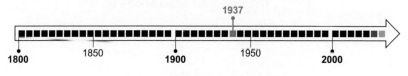

图 69.3　欧芬脑尔氧化反应的发现❷

---

❶ 麦尔外因-庞多夫-维利还原是欧芬脑尔氧化（Oppenauer oxidation）的逆反应。第 58 个反应机理已对此进行了介绍。

❷ 该反应大约于 1937 年被首次报道[69]。

70 臭氧分解
(ozonolysis)

$$70. \quad \underset{R}{\overset{R}{>}}=\underset{R}{\overset{R}{<}} + O_3 \longrightarrow \underset{R\ R}{\overset{R\ \ O-O\ \ R}{\diagup\diagdown}} \longrightarrow \underset{R}{\overset{O}{\|}}_R + \underset{R}{\overset{O}{\|}}_R$$

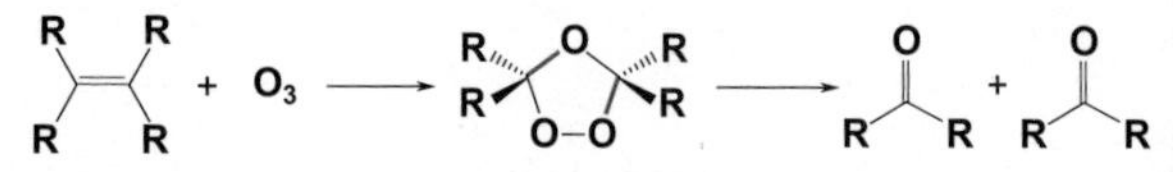

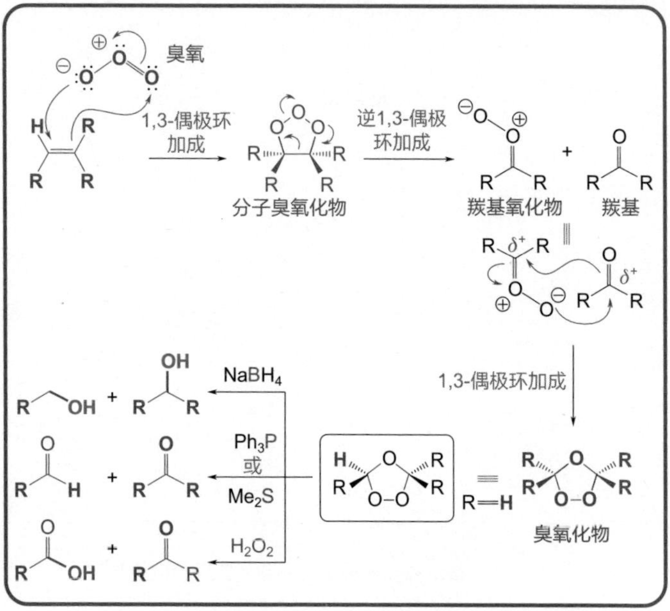

图 70.1 **臭氧分解的机理（克里奇反应的机理）**❶

---

❶[译者注：臭氧分解反应一般是指碳碳双键在臭氧作用下生成相应的羰基化合物的过程。]该反应的机理首先由克里奇（Criegee）[70a, 70b, 70c]提出，因此通常被称为克里奇机理（与第 29 个反应机理中介绍的克里奇氧化不同）。臭氧分解反应的第一步是 1,3-偶极环加成反应或（3 + 2）-环加成反应。

图 70.2 臭氧分解反应的替代条件 ❶

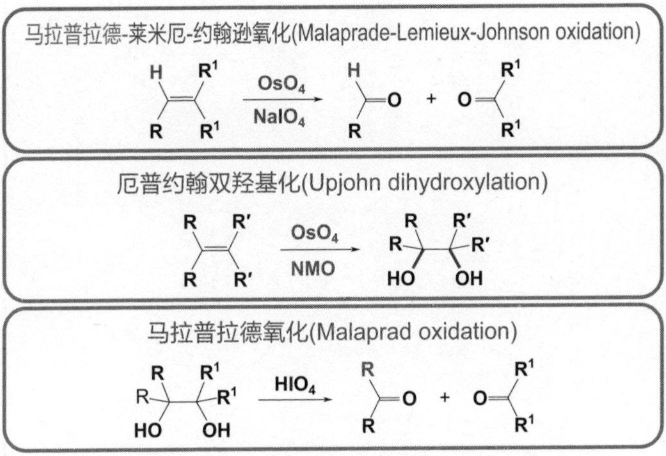

图 70.3 臭氧分解反应的相关反应 ❷

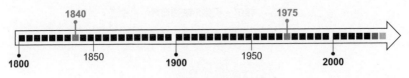

图 70.4 臭氧分解反应的发现 ❸

---

❶ 马拉普拉德 - 莱米厄 - 约翰逊试剂[70d]是使用臭氧[70e]的替代方法,随后通过 Ph_3P 或 Me_2S 生成醛和酮。莱米厄试剂[70f]也是臭氧的替代试剂,然后通过 H_2O_2 生成羧酸和酮。

❷ 厄普约翰双羟基化(参见第 93 个反应机理)及随后进行的马拉普拉德氧化(高碘酸氧化,参见第 29 个反应机理),也可用作臭氧分解反应的替代方法。

❸ 该反应大约于 1840 年[70g]被首次报道,相关的机理大约于 1975 年被首次提出[70b, 70c]。

71 帕尔－克诺尔合成
(Paal–Knorr syntheses)

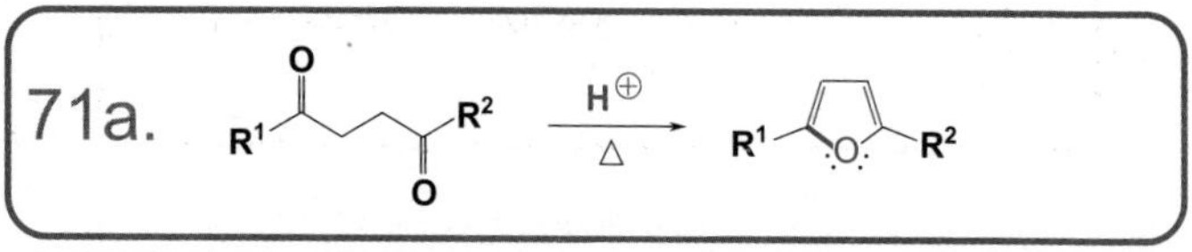

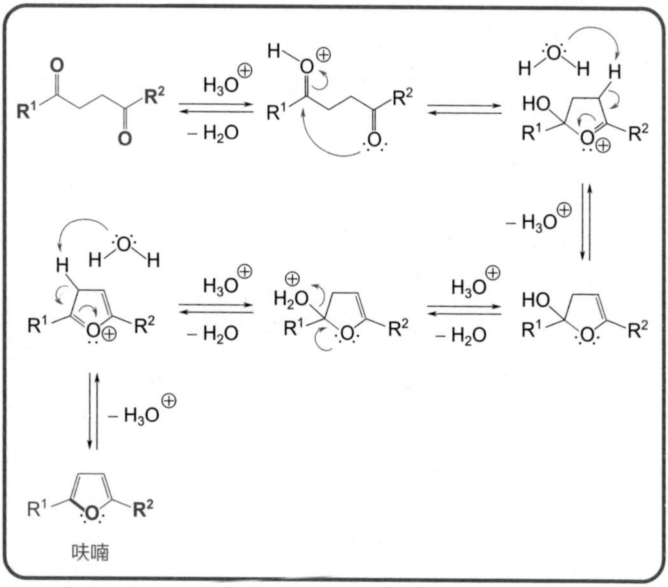

图 71.1　帕尔－克诺尔呋喃合成的机理 ❶

---

❶［译者注：帕尔-克诺尔合成是指由 1,4-二羰基化合物为原料环化制备呋喃、噻吩或吡咯类化合物的方法。］该反应最初被提出用于合成呋喃：帕尔-克诺尔呋喃合成（Paal-Knorr furan synthesis）。

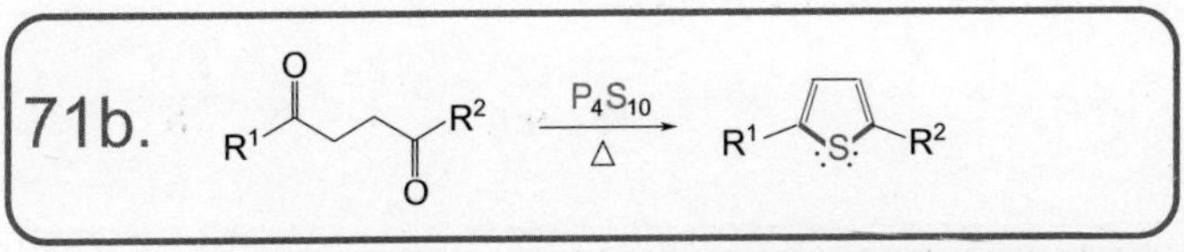

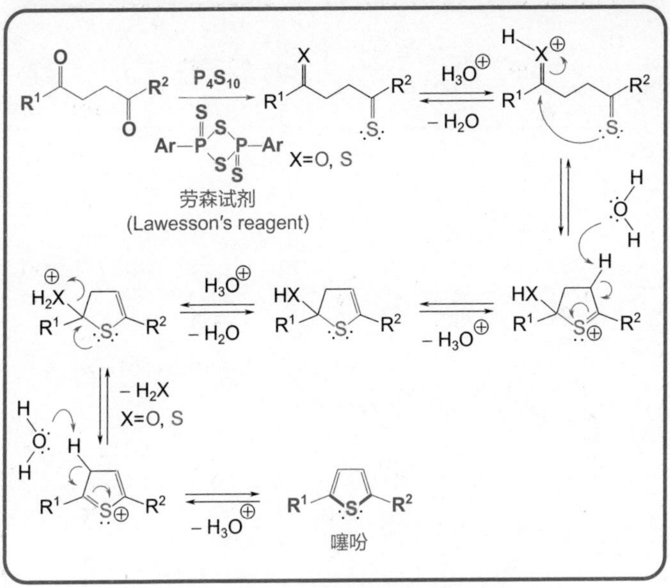

图 71.2　帕尔－克诺尔噻吩合成的机理 ❶

---

❶ 帕尔-克诺尔噻吩合成法是制备噻吩的有效方法，如使用劳森试剂[71a]。

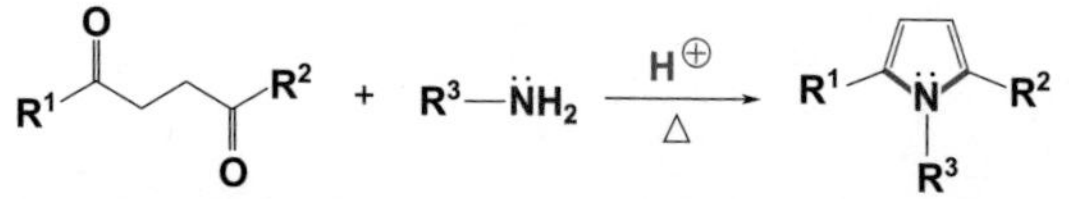

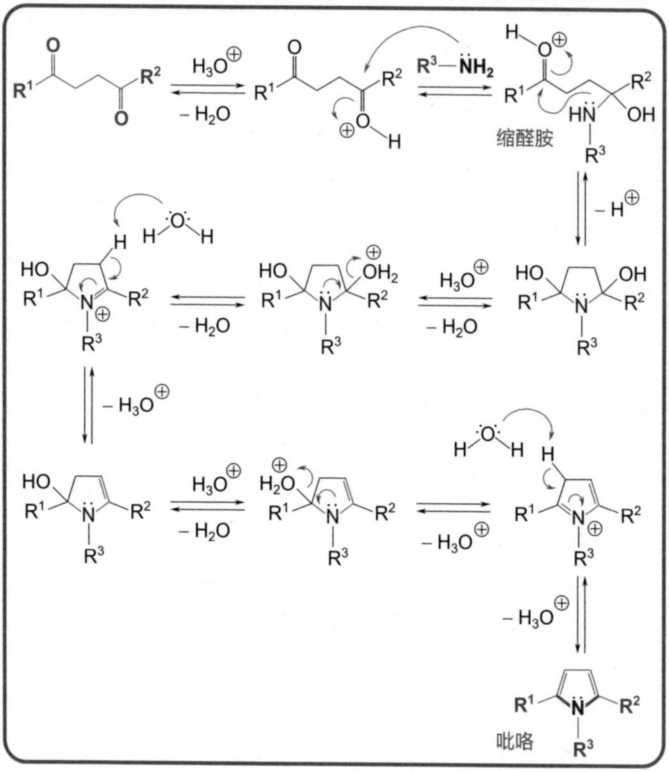

图 71.3　帕尔－克诺尔吡咯合成的反应机理 ❶

---

❶ 帕尔-克诺尔吡咯合成法最初被提出用于吡咯的合成。注意,不能将该反应与克诺尔吡咯合成反应相混淆。

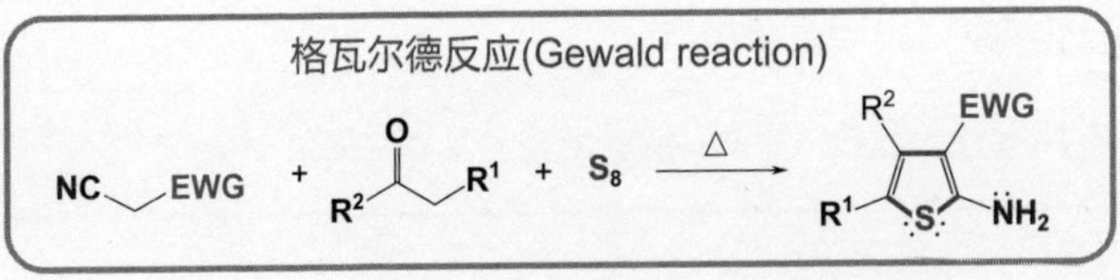

图 71.4 帕尔-克诺尔噻吩合成的相关反应 ❶

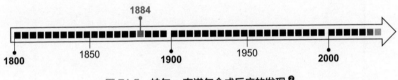

图 71.5 帕尔-克诺尔合成反应的发现 ❷

---

❶ 噻吩（2-氨基噻吩）可以通过格瓦尔缩合反应制得（参见第 41 个反应机理）。
❷ 该反应大约于 1884 年被首次报道[71b, 71c]。

72 帕特罗 – 布奇反应
(Paternò-Büchi reaction)

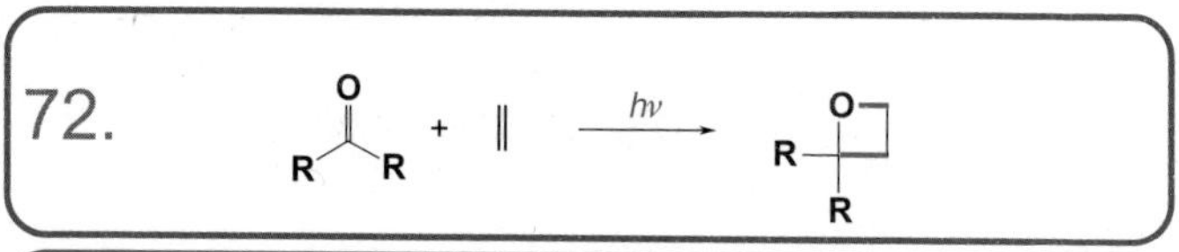

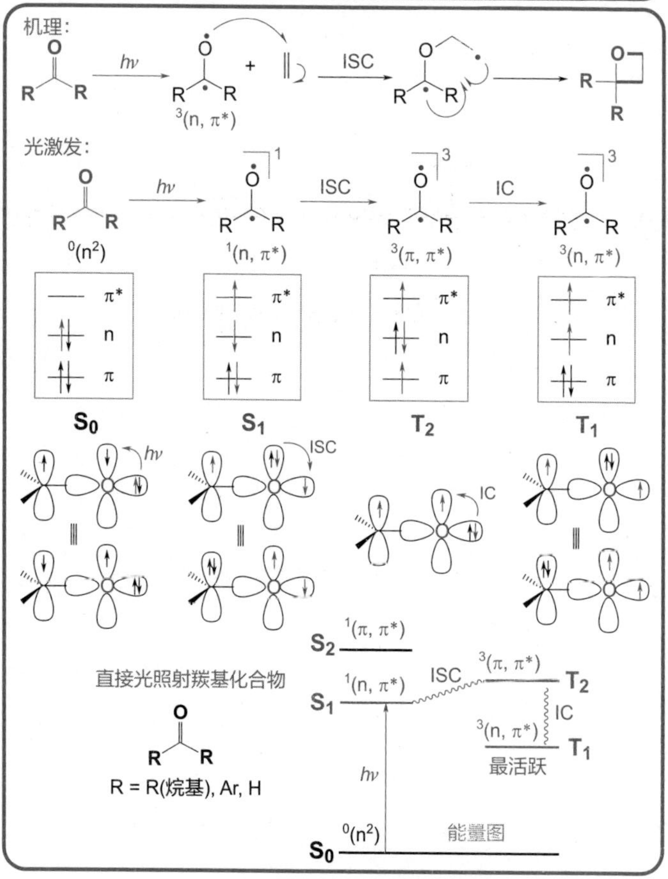

图 72.1 帕特罗 – 布奇反应的机理 ❶

❶ 帕特罗 - 布奇反应是指羰基化合物与烯烃的光化学 [$2_π+2_π$] 或 [2+2] - 环加成反应，利用其特殊的区域及立体选择性可以合成一系列氧杂环丁烷衍生物。根据伍德沃德 - 霍夫曼规则 (Woodward-Hoffmann rules)[64a, 64b, 64c]，该反应 ($4n=4π$) 不可在基态 [在热力学条件下 (△)] 发生，但可在激发态 [光照射条件下 (光催化)] 发生[2b]。

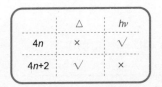

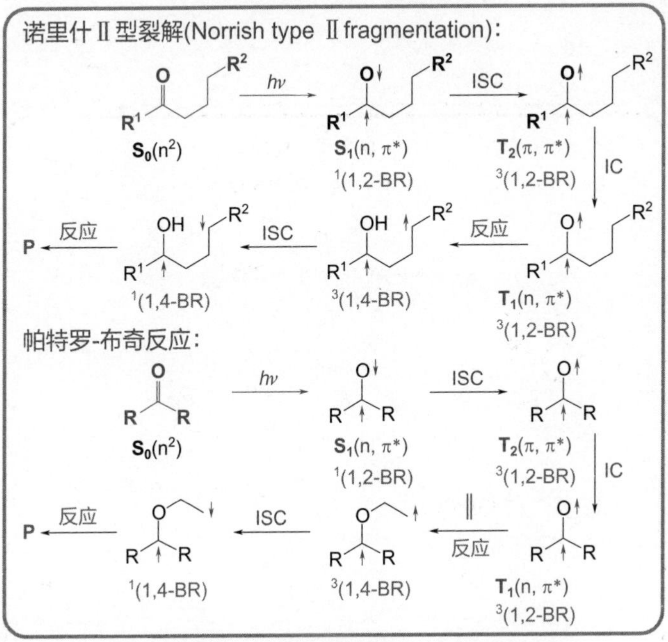

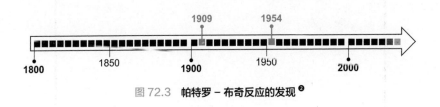

图72.2 诺里什 II 型反应与帕特罗 – 布奇反应机理 ❶

图72.3 帕特罗 – 布奇反应的发现 ❷

---

❶ 诺里什 II 型反应（参见第 67 个反应机理）与帕特罗 - 布奇环加成反应[2b]之间的机理相似性比较。

❷ 该反应大约是帕特罗于 1909 年[72a]以及布奇于 1954 年[72b]报道的。

73 保森-坎德反应
(Pauson-Khand reaction)

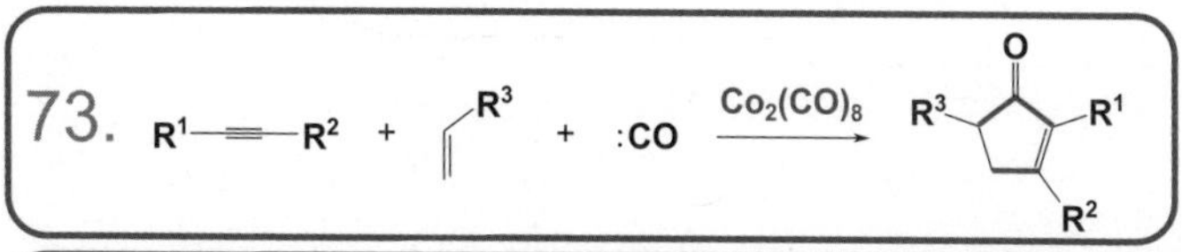

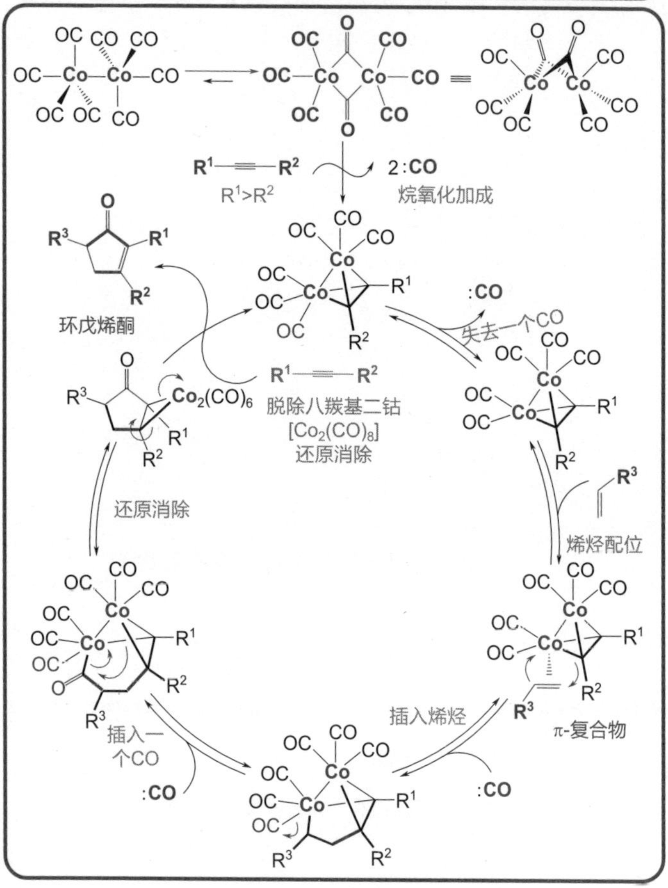

图 73.1 保森-坎德反应的机理 ❶

---

❶[译者注：保森-坎德反应是指八羰基二钴与烯烃、炔烃反应生成环戊烯酮衍生物的反应。]该反应是一种共催化的(2+2+1)-环加成反应。

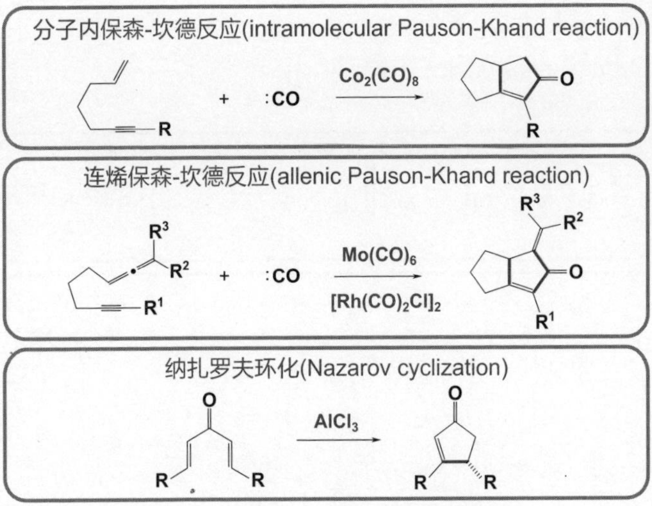

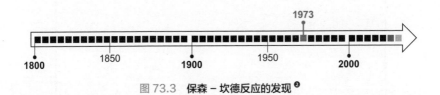

图 73.2 保森－坎德反应的变体反应 ❶

图 73.3 保森－坎德反应的发现 ❷

---

❶ 该反应包括几种变体反应：分子内的保森-坎德反应、连烯保森-坎德反应及其他反应[73a]。该反应也可由其他金属催化，如钼（Mo）、铑（Rh）等。纳扎罗夫环化经历了一种不同的[4π]顺旋电环化机理（参见第 64 个反应机理），但也会生成环戊烯酮。

❷ 该反应大约于 1973 年被首次报道[73b, 73c, 73d]。

74 肽（酰胺）偶联
[peptide (amide) coupling]

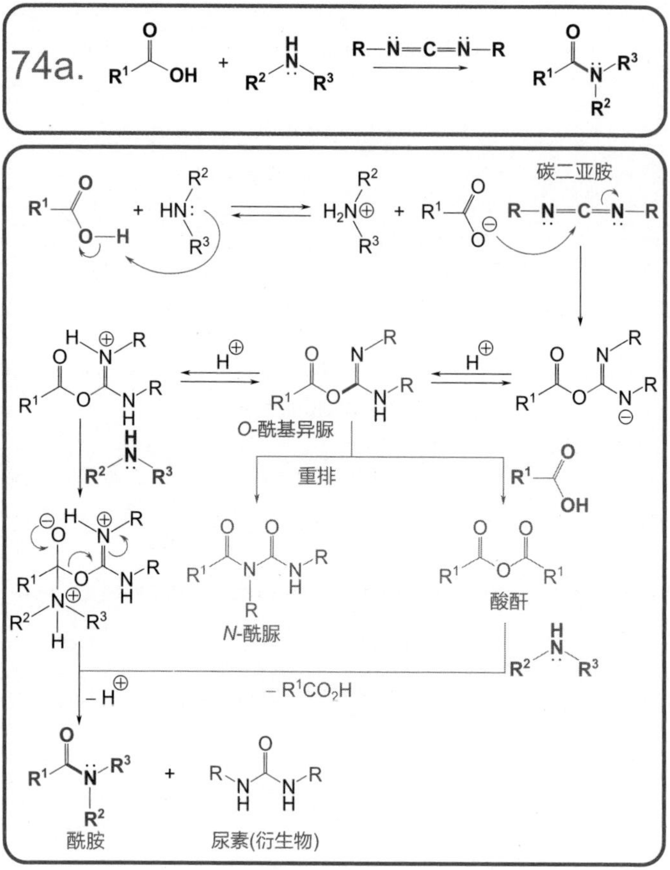

图 74.1 肽（酰胺）偶联反应的机理（DCC）[1]

---

[1] 基于碳二亚胺偶联试剂（DCC）的肽（酰胺）偶联反应 [peptide (amide) coupling] 的机理[74a, 74b]。

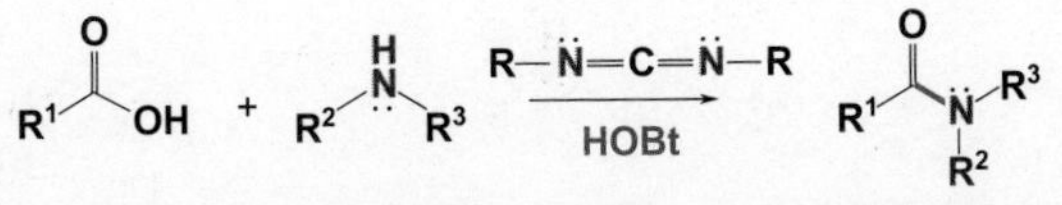

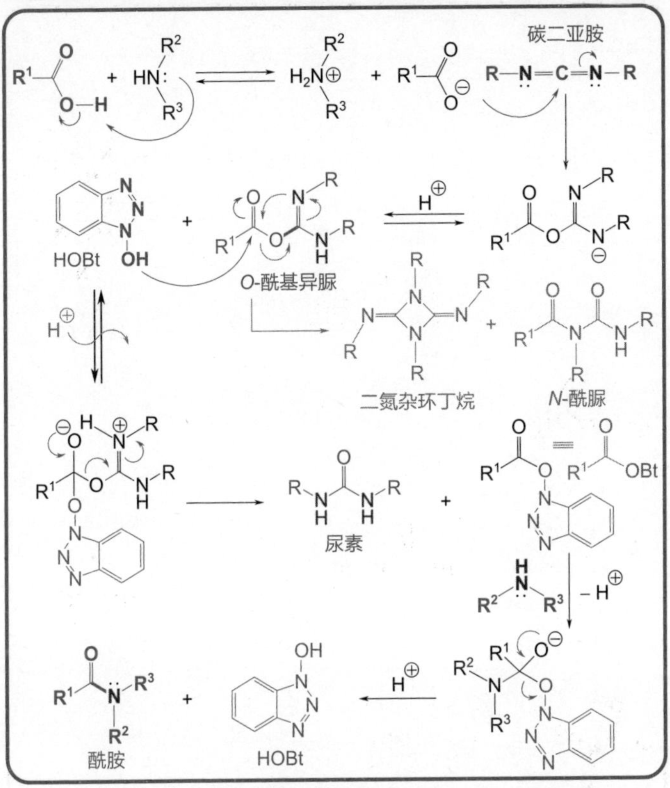

图 74.2 肽（酰胺）偶联反应的机理（DCC + HOBt）❶

---

❶ 基于碳二亚胺偶联剂和添加剂（DCC 和 HOBt）的肽（酰胺）偶联反应的机理[74a, 74b]。

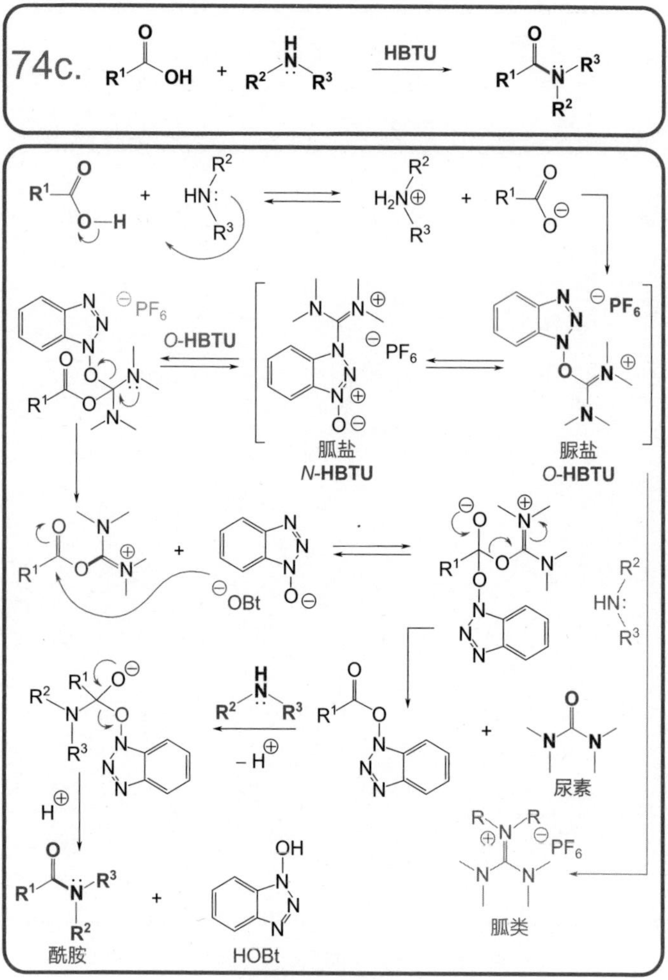

图 74.3　肽（酰胺）偶联的机理（HBTU）❶

---

❶ 基于苯并三唑=胍盐/脲盐偶联试剂（HBTU）的肽（酰胺）偶联的机理[74c]。

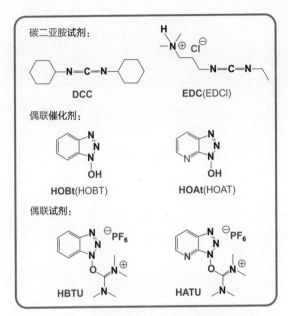

图 74.4　主要的肽（酰胺）偶联试剂和催化剂 ❶

图 74.5　肽（酰胺）偶联反应的发现 ❷

---

❶ 在酰胺偶联或肽合成中最常使用的试剂是碳二亚胺试剂（如 DCC[74d]、EDC[74e] 及许多其他试剂）；脒/脲盐 [HBTU[74f]、HATU[74g] 及许多鏻盐（PyBOP）[74h]]。在肽合成中最常使用的添加剂（催化剂）是 HOBt[74i] 和 HOAt 等。

❷ 肽（酰胺）偶联反应大约于 1901 年被首次报道[74j]；DCC 偶联试剂大约于 1955 年被首次报道[74k]；HBTU 偶联试剂大约于 1978 年被首次报道[74l]。

75 皮克特-斯彭格勒反应
(Pictet–Spengler reaction)

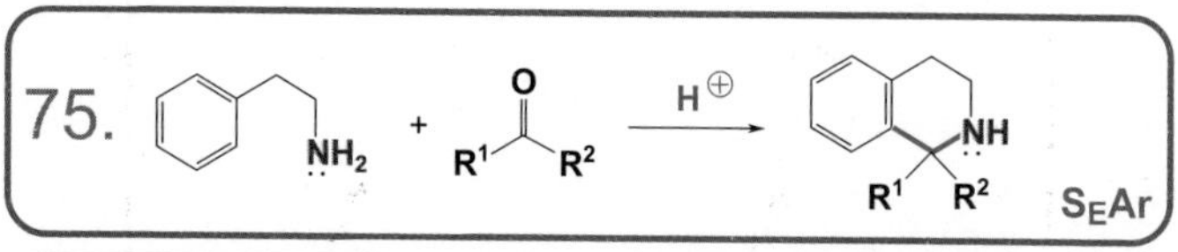

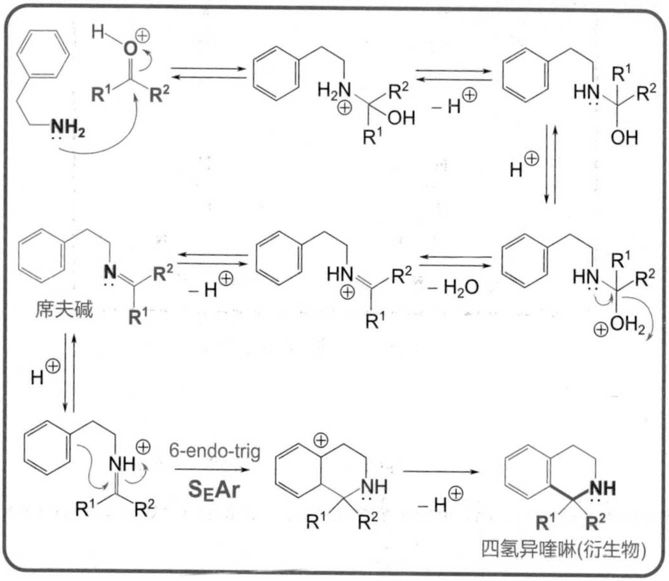

图 75.1 皮克特-斯彭格勒反应的机理 ❶

---

❶ [译者注：皮克特-斯彭格勒反应或皮克特-斯彭格勒缩合（Pictet-Spengler condensation）是指 β-芳基乙胺在酸性条件下与羰基化合物缩合再环化为 1,2,3,4-四氢异喹啉（衍生物）的反应。]其机理是曼尼希缩合（亚胺缩合）（席夫碱）（参见第 56 个反应机理）和芳香亲电取代（芳基正离子机理或 S_EAr）的组合。

75 皮克特 - 斯彭格勒反应

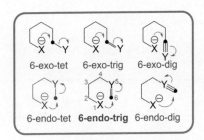

图 75.2　鲍德温规则（Baldwin's rules）❶

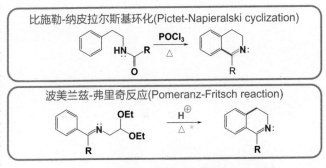

图 75.3　皮克特 – 斯彭格勒反应的相关反应 ❷

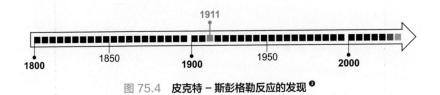

图 75.4　皮克特 – 斯彭格勒反应的发现 ❸

---

❶ 环化（S_EAr）步骤主要遵循鲍德温规则：6-endo-trig[75a]。
❷ 几种人名反应与皮克特 - 斯彭格勒反应有关：比施勒 - 纳皮拉尔斯基环化反应（参见第 19 个反应机理）、波美兰兹 - 弗里奇反应[19a, 19b]。这两种反应均生成异喹啉。
❸ 该反应大约于 1911 年被首次报道[75b]。

76 频哪醇－频哪酮重排
(pinacol-pinacolone rearrangement)

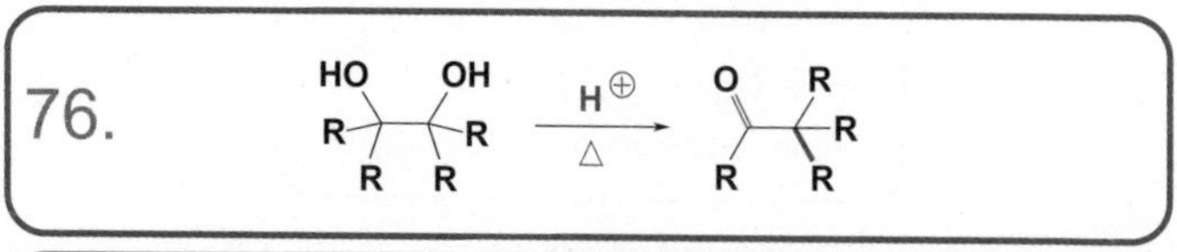

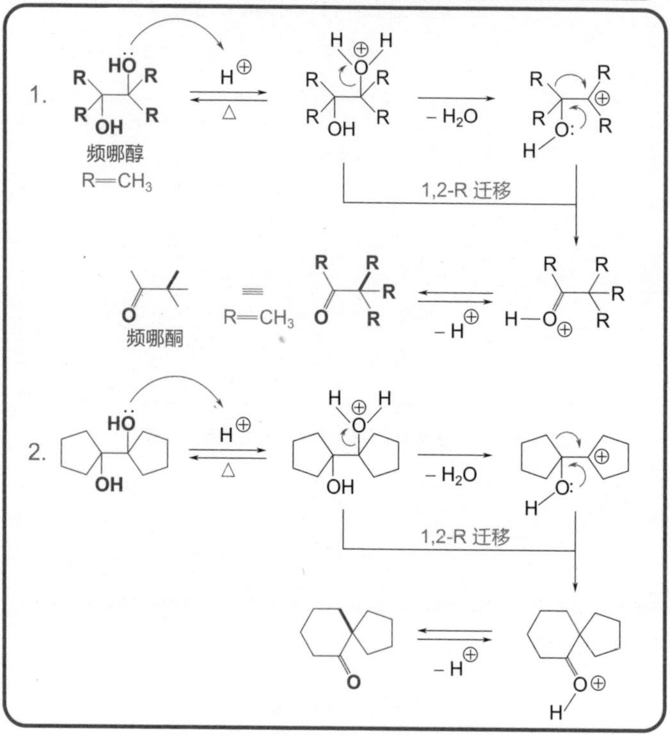

图 76.1 频哪醇－频哪酮重排的机理 ❶

---

❶ [译者注：频哪醇-频哪酮重排简称频哪醇重排，是指频哪醇在酸性条件下发生消除并重排生成不对称的酮，可用于螺环烃的合成。] 其机理与第 96 个反应机理中描述的瓦格纳-麦尔外因（Wagner-Meerwein）重排相关。应避免将频哪醇－频哪酮排与频哪醇偶联（参见第 57 个反应机理）相混淆。注意：2,3-二甲基-2,3-丁二醇称为频哪醇（pinacol），而 3,3-二甲基-2-丁酮称为频哪酮（pinacolone）。

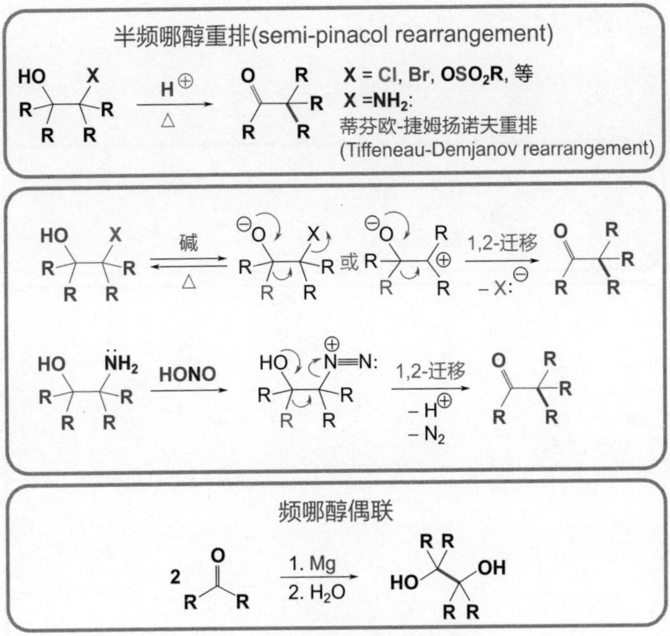

76 频哪醇-频哪酮重排

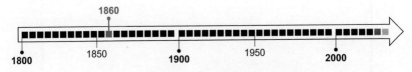

图 76.2　半频哪醇重排的机理 ❶

图 76.3　频哪醇－频哪酮重排反应的发现 ❷

---

❶ 半频哪醇重排的机理[1]与频哪醇重排类似，发生于 α- 取代醇中。如果 X=NH₂，则该反应称为蒂芬欧 - 捷姆扬诺夫重排[76a, 76b]。

❷ 该反应大约于 1860 年被首次报道[76c]。

77 波罗诺夫斯基反应
(Polonovski reaction)

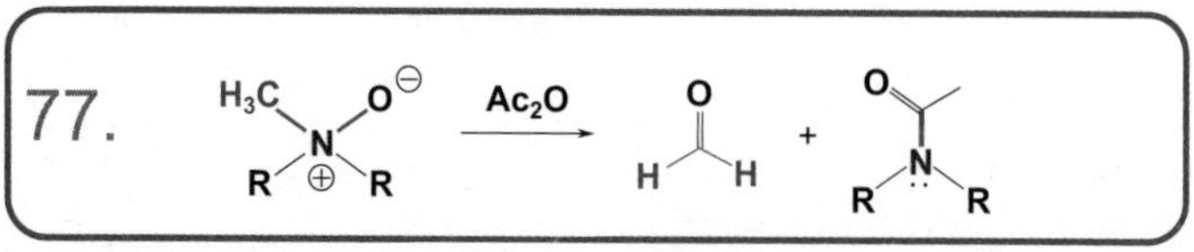

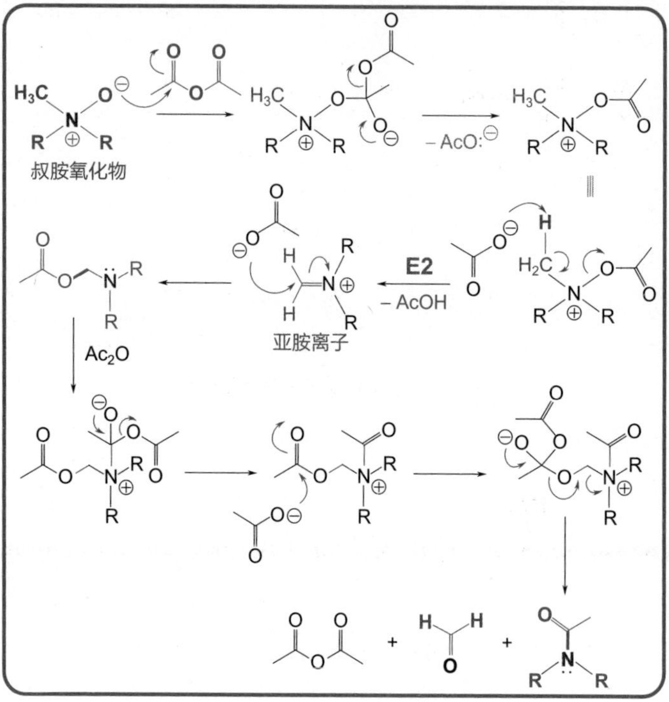

图 77.1　波罗诺夫斯基反应机理 ❶

---

❶ 波罗诺夫斯基反应也称为波罗诺夫斯基重排（Polonovski rearragement）[译者注：是指叔胺氧化物在醇溶液中与乙酸酐或乙酰氯反应，一个 C—N 键被断裂生成相应的乙酰胺衍生物和醛的反应]。该反应的关键中间体为亚胺离子（参见第 56 个反应机理中的曼尼希反应）。

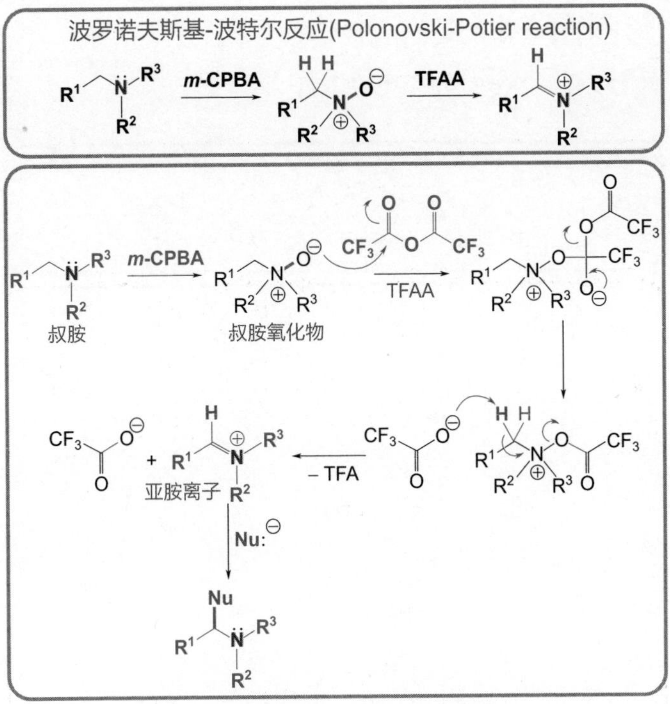

图 77.2　波罗诺夫斯基 – 波特尔反应的机理 ❶

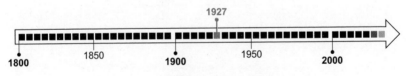

图 77.3　波罗诺夫斯基反应的发现 ❷

---

❶ 根据文献中的报道[77a, 77b]，波罗诺夫斯基 - 波特尔反应以三氟乙酸酐（TFAA）代替乙酸酐进行反应，并且亚胺离子可以被各种亲核试剂捕获。

❷ 该反应大约于 1927 年被首次报道[77c]。

78 普里莱扎耶夫环氧化
(Prilezhaev epoxidation)

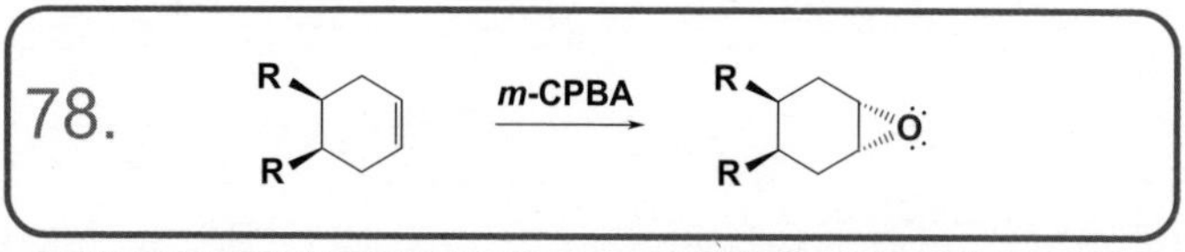

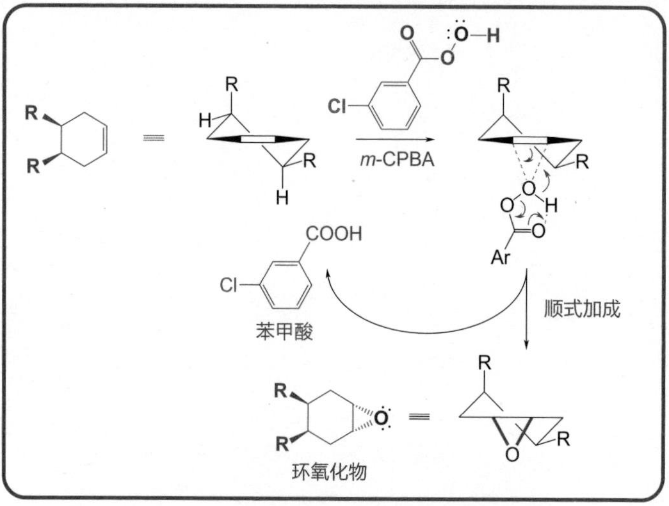

图 78.1 普里莱扎耶夫环氧化反应的机理 ❶

---

❶ 普里莱扎耶夫环氧化，也称为普里莱扎耶夫反应（Prilezhaev reaction）[译者注：是指烯烃在过氧化物作用下转化为环氧化物的反应]。

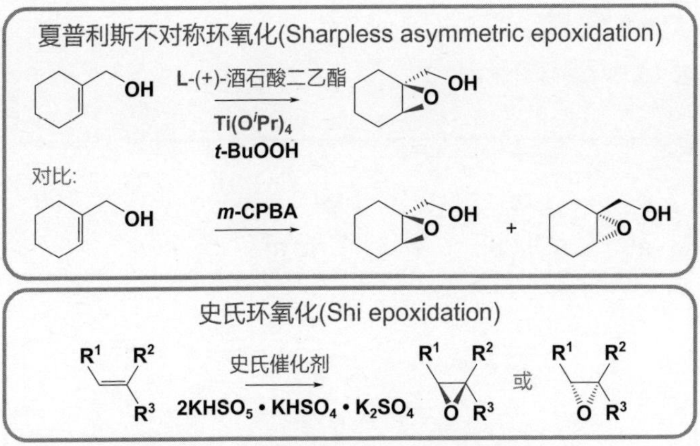

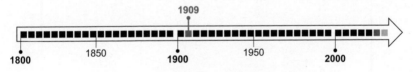

图 78.2 普里莱扎耶夫环氧化的相关反应 ❶

图 78.3 普里莱扎耶夫环氧化反应的发现 ❷

---

❶ 环氧化物的合成方法有很多,如夏普利斯不对称环氧化[78a]、生成对映体混合物的普里莱扎耶夫环氧化(Prilezhaev epoxidation)、史氏不对称环氧化(Shi asymmetric epoxidation)[78b],以及其他未举例的氧化反应[1]。

❷ 该反应大约于1909年被首次报道[78c]。

79 普林斯反应
(Prins reaction)

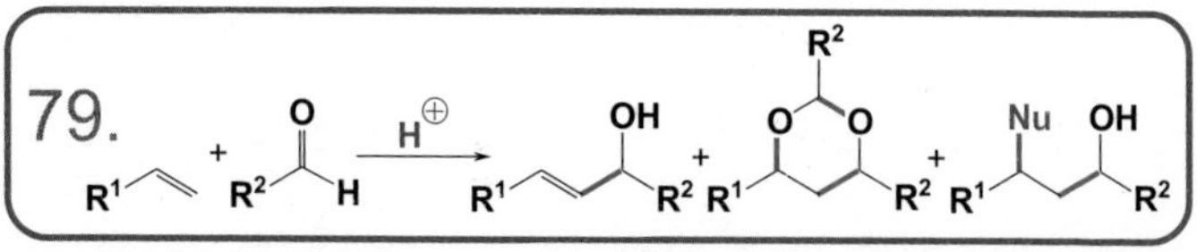

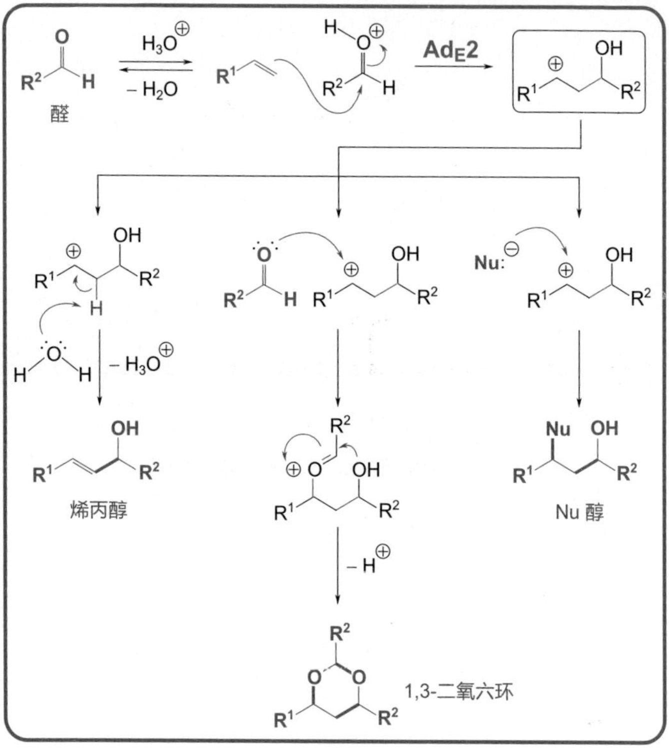

图 79.1　普林斯反应的机理 ❶

---

❶ 普林斯反应是一种质子化醛与烯烃加成缩合生成烯丙基醇的反应，在第 1 个反应机理亲电加成的实例中已有介绍。

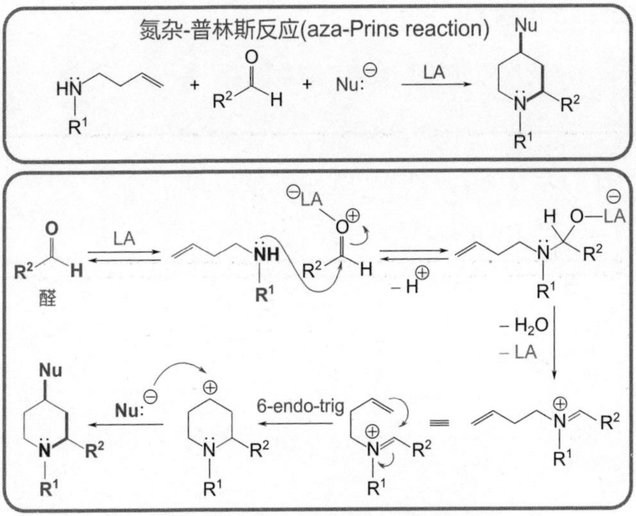

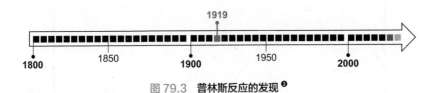

图 79.2　氮杂－普林斯反应的机理 ❶

图 79.3　普林斯反应的发现 ❷

---

❶ 氮杂-普林斯反应的机理与普林斯反应相关[79a, 79b]。该反应除了生成哌啶结构（请参见第 75 个反应机理中提到的鲍德温规则：6-endo-trig），还存在其他变体，如普林斯-频哪醇反应（Prins-pinacol reaction）[79c]。

❷ 该反应大约于 1919 年被首次报道[79d, 79e]。

80 普默勒重排
(Pummerer rearrangement)

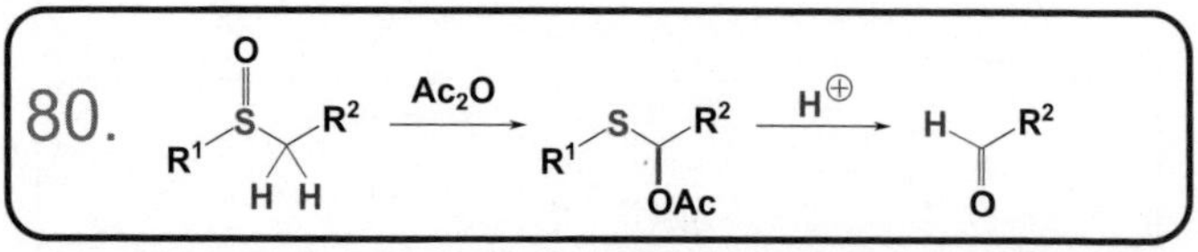

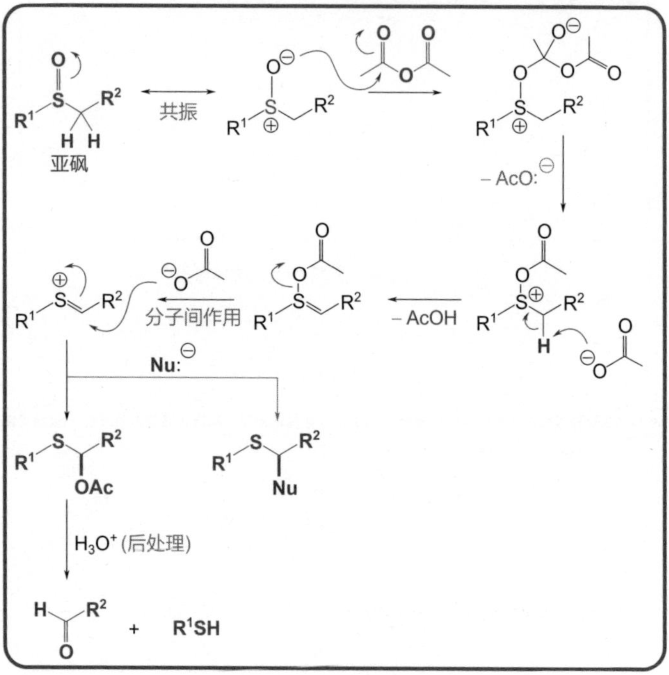

图 80.1 普默勒重排的机理 ❶

---

❶ 普默勒重排，也称为普默勒裂解（Pummerer fragmentation）[译者注：是指亚砜衍生物在乙酸酐作用下重排为 α-酰氧基硫醚的反应，其中硫原子被还原，而邻位碳被氧化]。

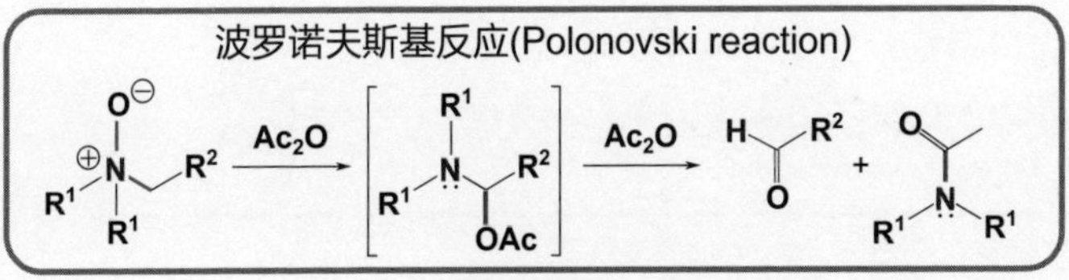

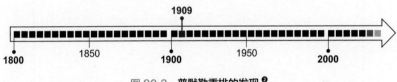

图 80.2 普默勒重排的相关反应 ❶

图 80.3 普默勒重排的发现 ❷

---

❶ 波罗诺夫斯基反应的机理(参见第 77 个反应机理)与普默勒重排有关。波罗诺夫斯基反应中的氧化胺与普默勒重排反应中的亚砜具有类似的作用。

❷ 该反应大约于 1909 年被首次报道[80]。

81 兰伯格－贝克伦德重排
(Ramberg-Bäcklund rearrangement)

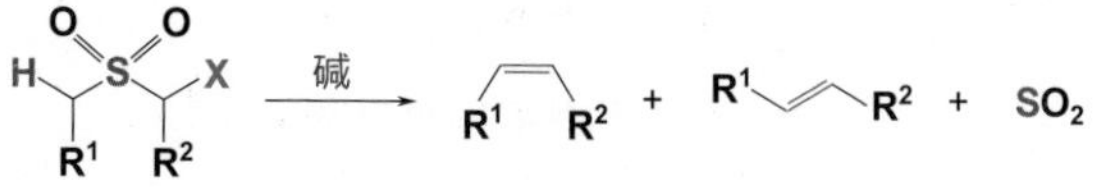

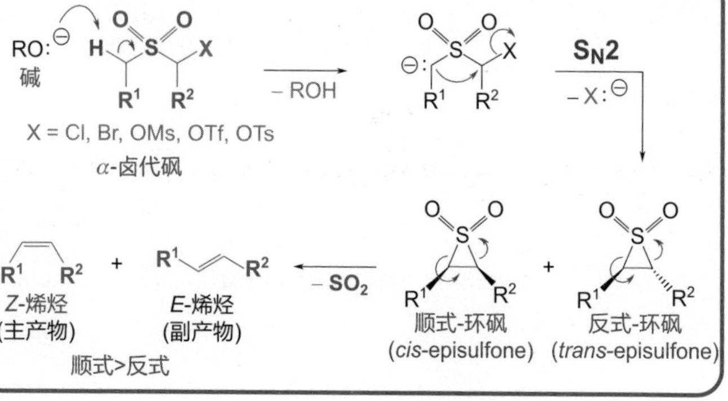

图 81.1 兰伯格－贝克伦德重排的机理 ❶

---

❶ [译者注：兰伯格-贝克伦德重排，也称为兰伯格-贝克伦德反应（Ramberg-Bäcklund reaction），是指 α-卤代砜在碱的引发下经由一个三元环砜中间体重排为烯的反应。]其机理是第 2 个反应机理中所述的双分子亲核取代（S_N2）和随后的协同消除（螯合消除）反应[1a, 81a]。

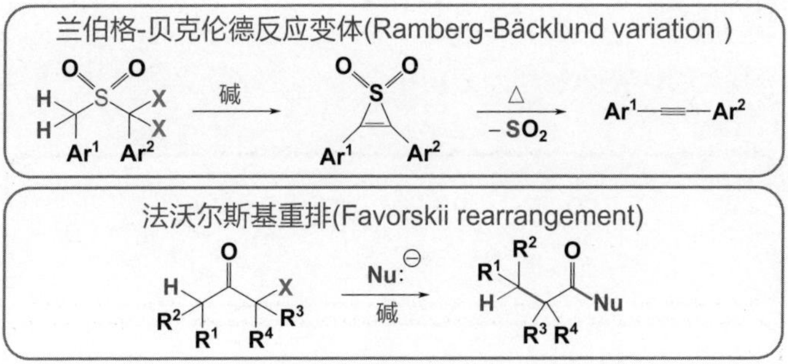

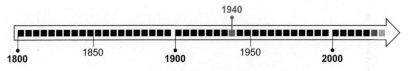

图 81.2 兰伯格－贝克伦德重排的相关反应 ❶

图 81.3 兰伯格－贝克伦德重排的发现 ❷

---

❶ 兰伯格 - 贝克伦德重排包括多种变体，如形成炔烃的反应[81b, 1a]。法沃尔斯基重排中的 S_N2 步骤（参见第 37 个反应机理）与兰伯格 - 贝克伦德重排有关。
❷ 该反应大约于 1940 年被首次报道[81c]。

82 雷福尔马茨基反应
(Reformatsky reaction)

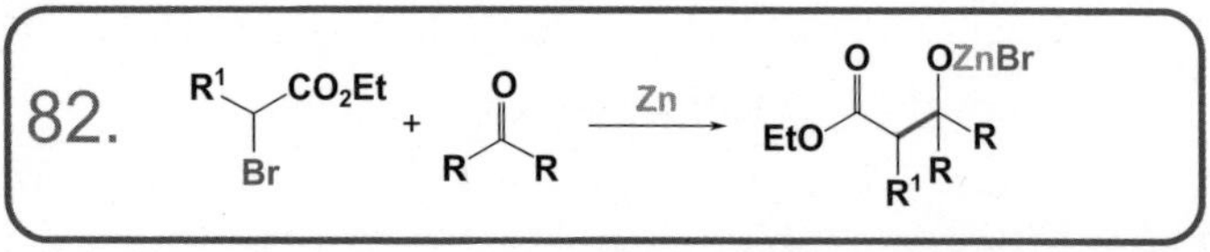

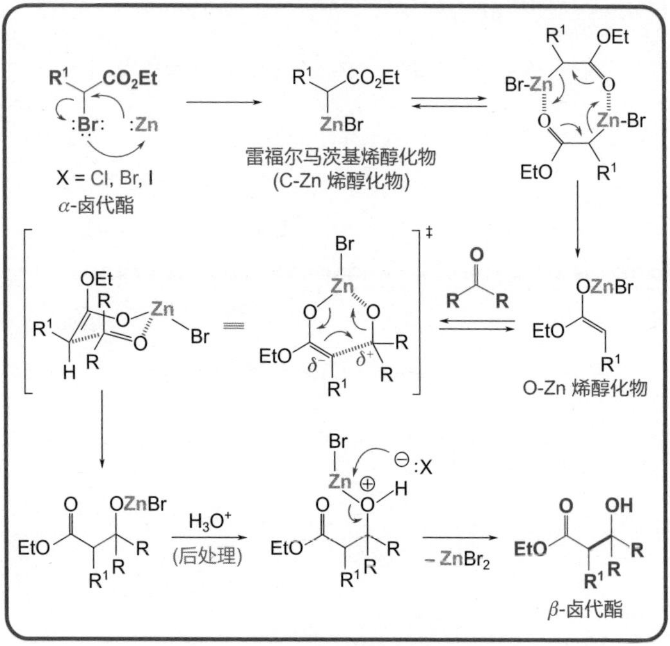

图 82.1 **雷福尔马茨基反应的机理** ❶

---

❶ 雷福尔马茨基反应，也可写作 Reformatskii reaction）[译者注：是指 α- 卤代酯与醛或酮在金属锌和酸作用下转化为 β- 羟基酯的反应]。该反应是一种缩合反应，其机理类似于羟醛缩合反应（参见第 83 个反应机理）。

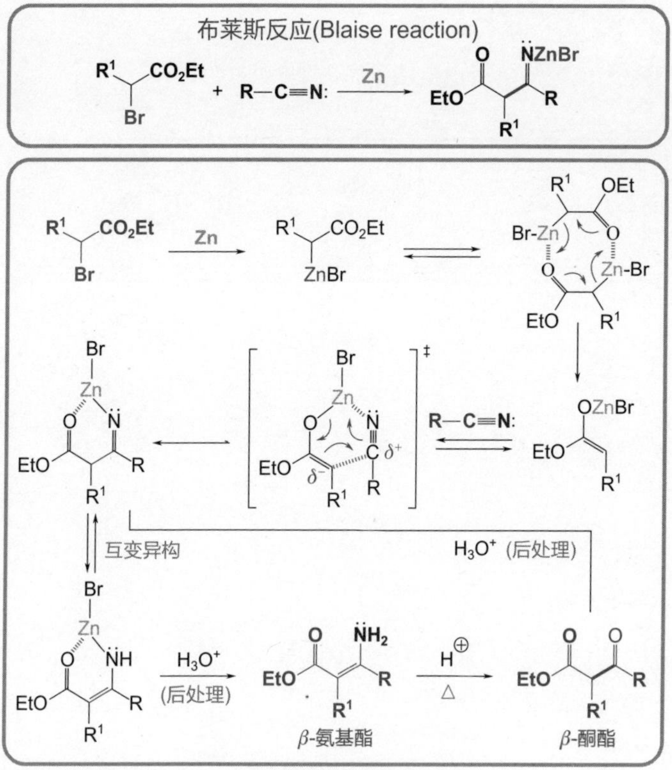

图 82.2 **布莱斯反应的机理** ❶

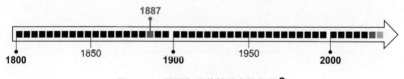

图 82.3 **雷福尔马茨基反应的发现** ❷

---

❶ 布莱斯反应是雷福尔马茨基反应的一种变体反应[82a, 82b]。在该情况下,预先形成的雷福尔马茨基烯醇化物(C-Zn 或 O-Zn 烯醇化物)与氰基反应,而不是与醛或酮反应。

❷ 该反应大约于 1887 年被首次报道[82]。

83 罗宾逊环化
(Robinson annulation)

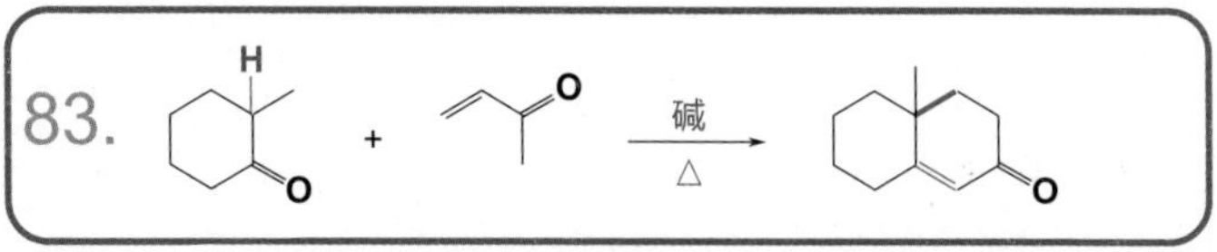

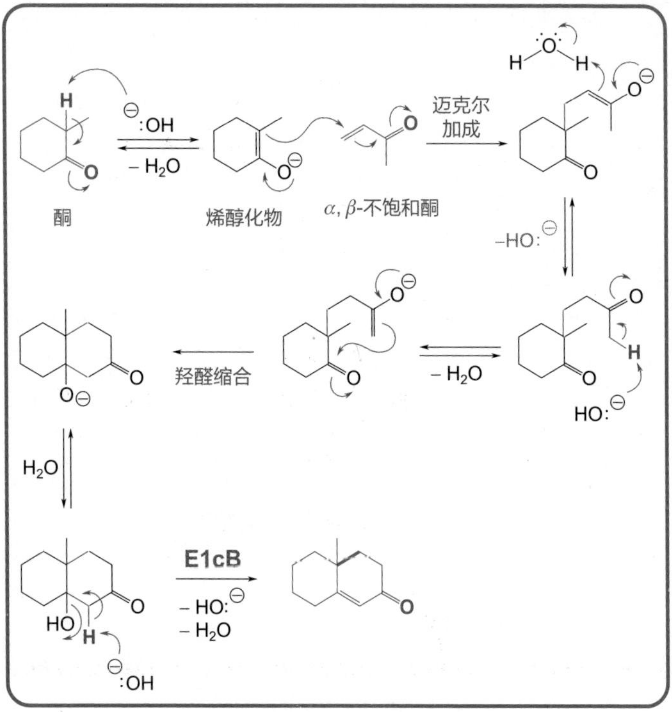

图 83.1　罗宾逊环化反应的机理 ❶

---

❶［译者注：罗宾逊环化是指通过分子内羟醛缩合生成环己酮衍生物的反应。］该反应的机理首先经过迈克尔共轭加成反应［Michael conjugate addition（参见第 59 个反应机理）］，随后进行羟醛缩合，最后发生 **E1cB** 消除（参见第 6 个反应机理）。

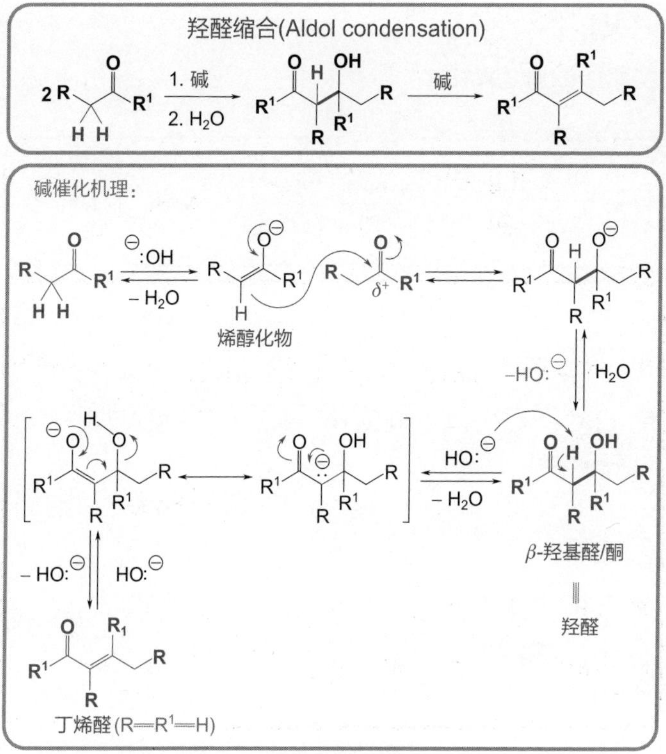

83 罗宾逊环化

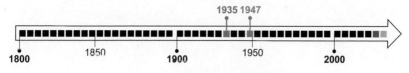

图 83.2　羟醛缩合机理 ❶

图 83.3　罗宾逊环化反应的发现 ❷

---

❶ 羟醛缩合(也称为醇醛缩合)反应可在碱催化下生成 β-羟基醛(醇醛)或酮。形成的醇醛经过消除并生成丁烯醛(巴豆醛)(巴豆缩合)[1]。

❷ 该反应大约于 1935 年被首次报道[83a]。1947 年, Robert Robinson 因其在生物碱领域的成就获得了诺贝尔化学奖[83b]。

84 夏皮罗反应
(Shapiro reaction)

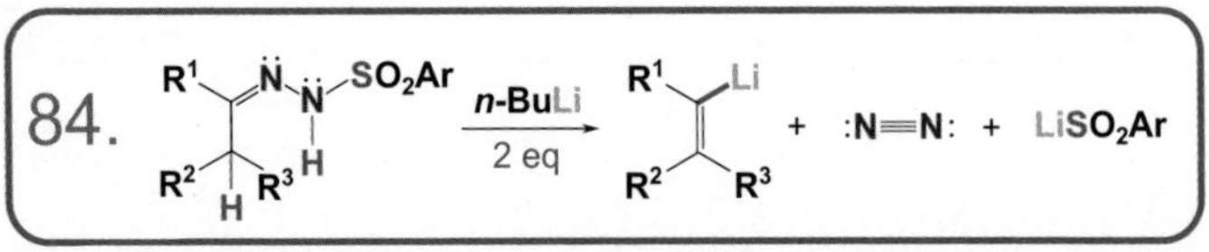

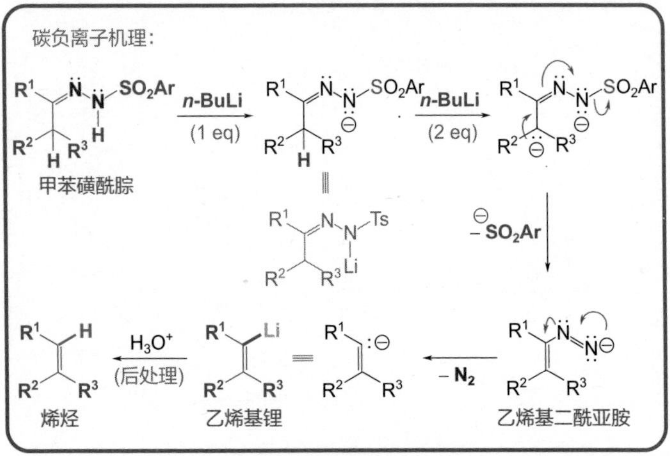

图 84.1 夏皮罗反应的机理 ❶

---

❶[译者注：夏皮罗反应是指醛或酮的对甲苯磺酰腙在强碱（如正丁基锂）作用下发生消除生成烯烃的反应。]该反应是一种基于碳负离子机理的消除反应。

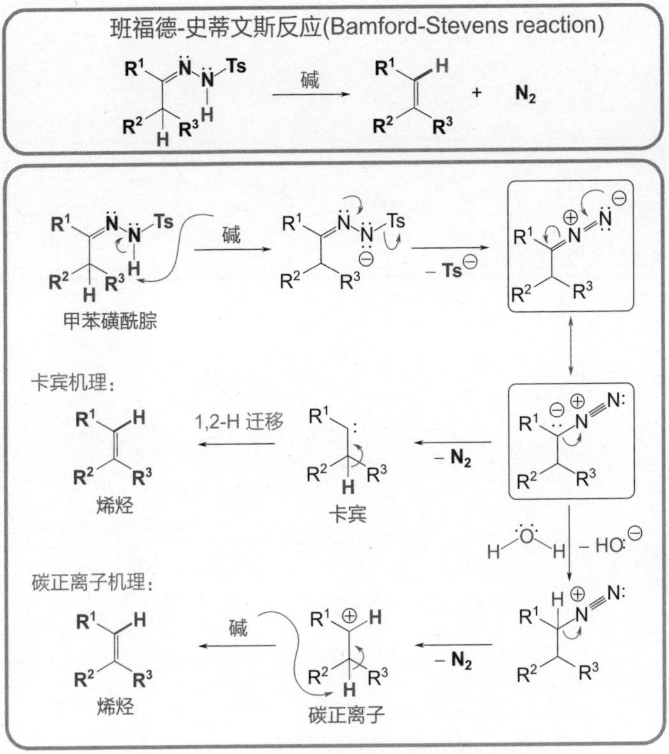

84 夏皮罗反应

班福德-史蒂文斯反应(Bamford-Stevens reaction)

图 84.2　班福德 – 史蒂文斯反应的机理 ❶

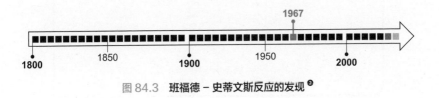

图 84.3　班福德 – 史蒂文斯反应的发现 ❷

---

❶ 班福德 - 史蒂文斯反应是夏皮罗反应的常见变体反应，可能涉及卡宾和碳正离子两种机理[84a]。

❷ 该反应大约于 1967 年被首次报道[84b]，具体参见文献[84c, 84d]。

85 薗头交叉偶联反应
(Sonogashira cross coupling)

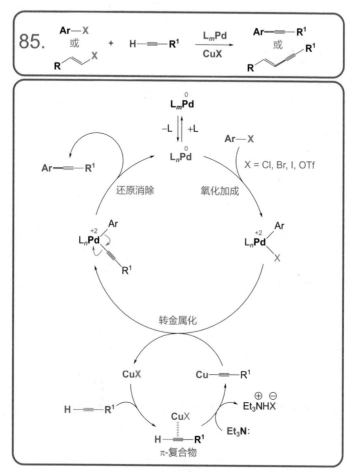

图 85.1　薗头交叉偶联反应的机理 ❶

---

❶ 薗头交叉偶联反应是一种 Pd-Cu 催化下的卤代烃和端基炔烃之间的交叉偶联反应。图中展示了一种简化易懂的 Pd-Cu 催化循环的一般机理。

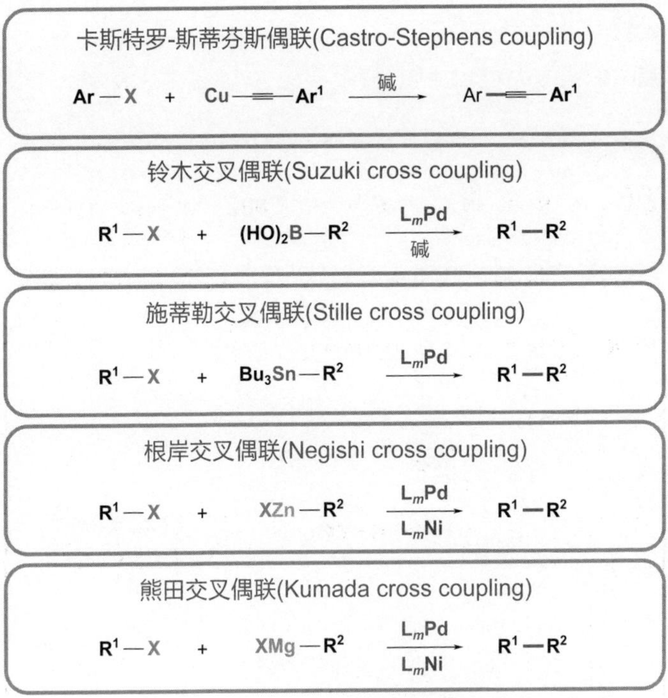

卡斯特罗-斯蒂芬斯偶联(Castro-Stephens coupling)

$$Ar-X + Cu-\!\!\!\equiv\!\!\!-Ar^1 \xrightarrow{\text{碱}} Ar-\!\!\!\equiv\!\!\!-Ar^1$$

铃木交叉偶联(Suzuki cross coupling)

$$R^1-X + (HO)_2B-R^2 \xrightarrow[\text{碱}]{L_mPd} R^1-R^2$$

施蒂勒交叉偶联(Stille cross coupling)

$$R^1-X + Bu_3Sn-R^2 \xrightarrow{L_mPd} R^1-R^2$$

根岸交叉偶联(Negishi cross coupling)

$$R^1-X + XZn-R^2 \xrightarrow[L_mNi]{L_mPd} R^1-R^2$$

熊田交叉偶联(Kumada cross coupling)

$$R^1-X + XMg-R^2 \xrightarrow[L_mNi]{L_mPd} R^1-R^2$$

图 85.2　薗头交叉偶联的相关反应 ❶

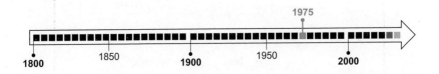

图 85.3　薗头交叉偶联反应的发现 ❷

---

❶ 铜催化的卡斯特罗-斯蒂芬斯交叉偶联中芳基卤化物和铜(Ⅰ)乙酰化物(预先形成或原位生成)形成 C—C 键[85a]。铃木交叉偶联(参见第 89 个反应机理)、施蒂勒交叉偶联(参见第 88 个反应机理)、根岸交叉偶联(参见第 66 个反应机理)和熊田交叉偶联(参见第 53 个反应机理)等也与薗头交叉偶联有关。

❷ 该反应大约于 1975 年被首次报道[85b]。

86 施陶丁格反应
(Staudinger reaction)

$$86.\ R-\ddot{N}=\overset{\oplus}{N}=\overset{\ominus}{\ddot{N}}: + Ph_3P: \longrightarrow R-\ddot{N}H_2 + :N\equiv N: + O=PPh_3$$

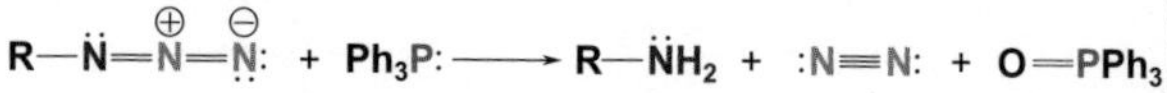

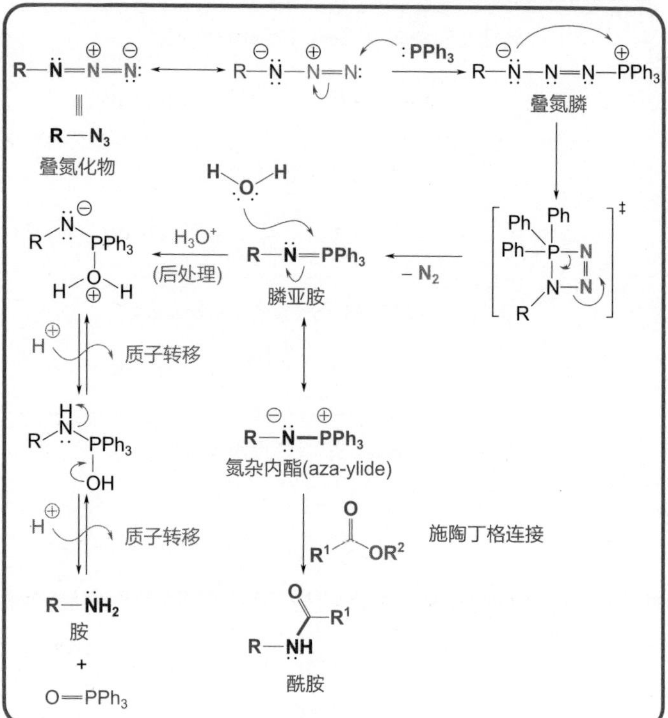

图 86.1 施陶丁格反应机理 ❶

---

❶ 施陶丁格反应，也称为施陶丁格还原（Staudinger reduction），是一种还原反应，通过使用三苯基膦将叠氮化物还原成伯胺。注意，应避免将其与施陶丁格合成（Staudinger synthesis）、施陶丁格烯酮环加成反应（Staudinger ketene cycloaddition reaction）相混淆，如 β-内酰胺的形成[86a, 86b]。

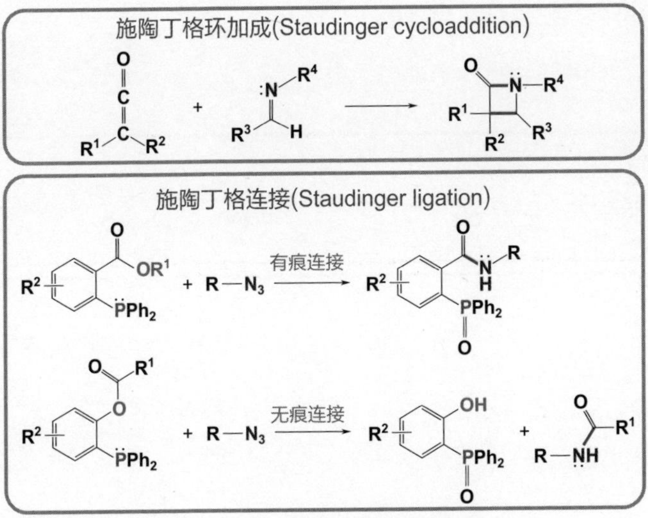

图 86.2 施陶丁格环加成及连接反应 ❶

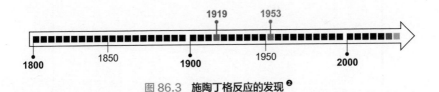

图 86.3 施陶丁格反应的发现 ❷

---

❶ 施陶丁格连接反应[86c, 86d]是施陶丁格反应的一种改进。施陶丁格连接生成的氮杂叶立德被酯捕获,形成酰胺键。反应一般包括有痕和无痕两种连接类型[86e]。

❷ 该反应大约于 1919 年被首次报道[86f]。1953 年,Hermann Staudinger 凭借化学大分子研究的有关成就获得了诺贝尔化学奖[86g]。

87 斯特格里奇酯化
(Steglich esterification)

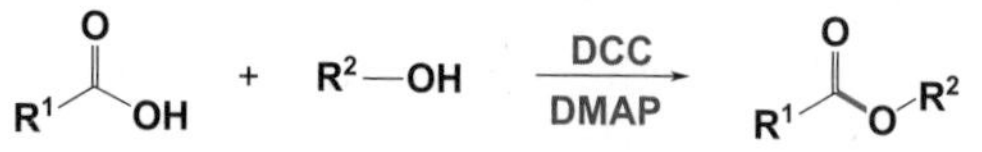

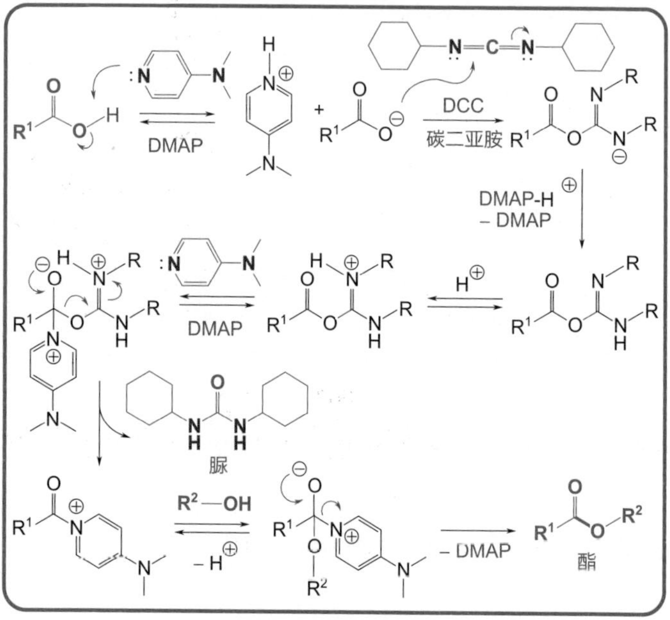

图 87.1 斯特格里奇酯化的机理（DCC + DMAP）❶

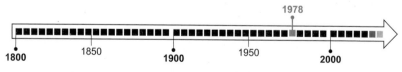

图 87.2 斯特格里奇酯化反应的发现 ❶

---

❶[译者注：斯特格里奇酯化是一种酯偶联反应，具体是指以 DCC 为偶联试剂，DMAP 为催化剂促进的酯化反应。]与肽（酰胺）偶联机理（参见第 74 个反应机理）或费歇尔酯化（Fischer esterification）相比，该反应机理涉及碳二亚胺偶联剂和 DMAP 催化剂的使用[87a]。

❷斯特格里奇酯化反应大约于 1978 年被首次报道[87b]。

87 斯特格里奇酯化

87b.

$R^1-COOH + R^2-OH \xrightarrow[\text{DMAP}]{\text{HOBt}} R^1-COO-R^2$

(R-N=C=N-R)

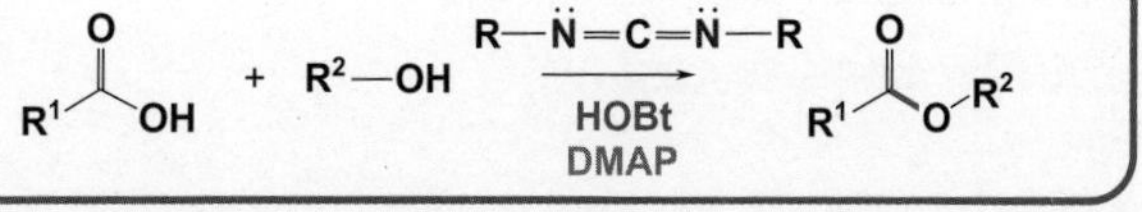

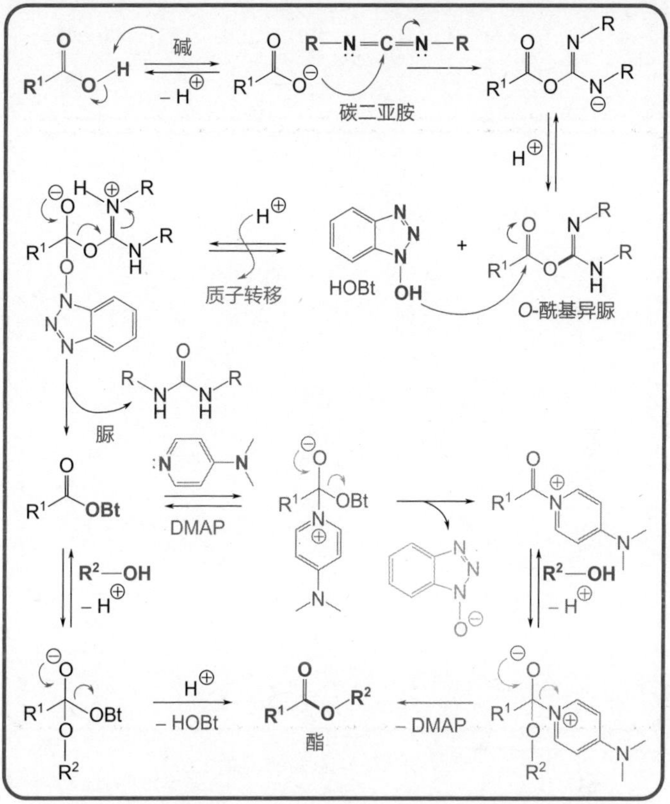

图 87.3 斯特格里奇酯化反应的机理（DCC + HOBt + DMAP）❶

中间步骤标注：碱、碳二亚胺、质子转移、HOBt、O-酰基异脲、脲、DMAP、酯

---

❶ 可在不添加 DMAP 催化剂的情况下，使用 DCC 和其他肽（酰胺）偶联试剂（例如 HOBt）进行斯特格里奇酯化[87a]。

88 施蒂勒交叉偶联反应
(Stille cross coupling)

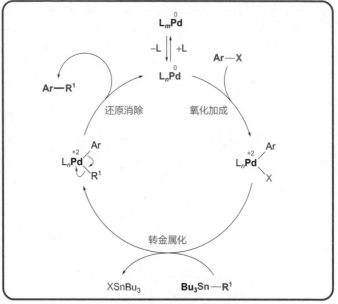

图 88.1　施蒂勒交叉偶联反应的机理 ❶

---

❶ 施蒂勒交叉偶联和右田 - 小杉 - 施蒂勒交叉偶联（Migita-Kosugi-Stille cross coupling）都是在钯催化下使用有机锡试剂和芳基卤化物或其他亲电试剂进行 C—C 键偶联的一种通用反应。上图为简化的反应机理图。

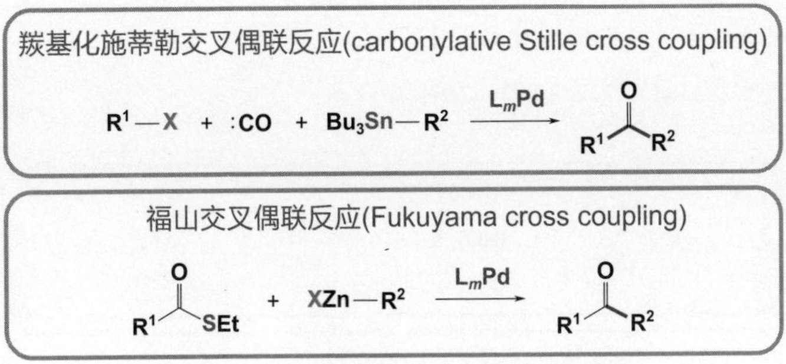

羰基化施蒂勒交叉偶联反应(carbonylative Stille cross coupling)

$$R^1-X + :CO + Bu_3Sn-R^2 \xrightarrow{L_mPd} R^1\overset{O}{\underset{}{-}}R^2$$

福山交叉偶联反应(Fukuyama cross coupling)

$$R^1\overset{O}{\underset{}{-}}SEt + XZn-R^2 \xrightarrow{L_mPd} R^1\overset{O}{\underset{}{-}}R^2$$

图 88.2　施蒂勒交叉偶联反应的相关反应 ❶

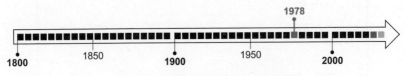

图 88.3　施蒂勒交叉偶联反应的发现 ❷

---

❶ 羰基化施蒂勒交叉偶联与施蒂勒交叉偶联有关，通过使用芳基卤化物或其他亲电试剂、有机锡和一氧化碳形成两个 C—C 键，生成酮[88a]。除此之外，福山交叉偶联通过使用硫酯和有机锌化合物形成 C—C 键，也可以生成酮[88b]；或者借助利贝斯金德 - 斯罗格尔交叉偶联 [Liebeskind-Srogl cross coupling（参见第 55 个反应机理）] 通过使用硫酯和硼酸形成 C—C 键，进而生成酮。

❷ 该反应大约于 1978 年被首次报道。

89 铃木交叉偶联反应
(Suzuki cross coupling)

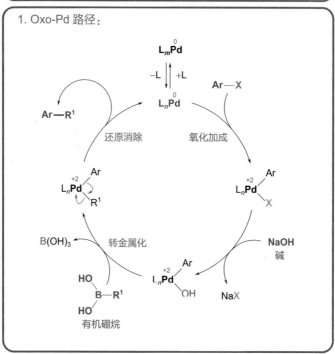

图 89.1　铃木交叉偶联的机理 [Oxo-Pd 路径] ❶

---

❶ 铃木交叉偶联即铃木 - 宫浦交叉偶联（Suzuki-Miyaura cross coupling），是一种钯催化的交叉偶联反应（使用芳基卤化物和有机硼酸形成碳碳键）。它是有机化学和药物化学合成中最重要的反应之一，而 Oxo-Pd 路径是较好的反应机理[89a]。

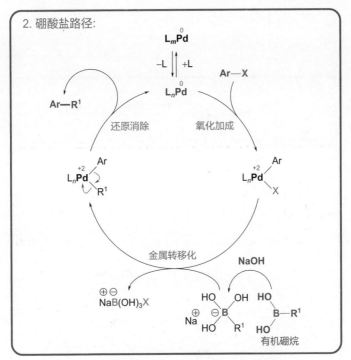

图 89.2　铃木交叉偶联反应的机理 [硼酸盐路径] ❶

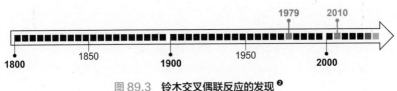

图 89.3　铃木交叉偶联反应的发现 ❷

---

❶ 该反应机理也可以通过硼酸盐路径来解释。上图显示了一种简化的一般机理[89b]。

❷ 该反应大约于 1979 年被首次报道[89c, 89d]。鉴于 Azira Suzuki、Richard F. Heck 和 Ei-ichi Negishi 对 Pd 催化交叉偶联反应的研究成就,三人共同获得了 2010 年诺贝尔化学奖[46c]。

90 斯文氧化
(Swern oxidation)

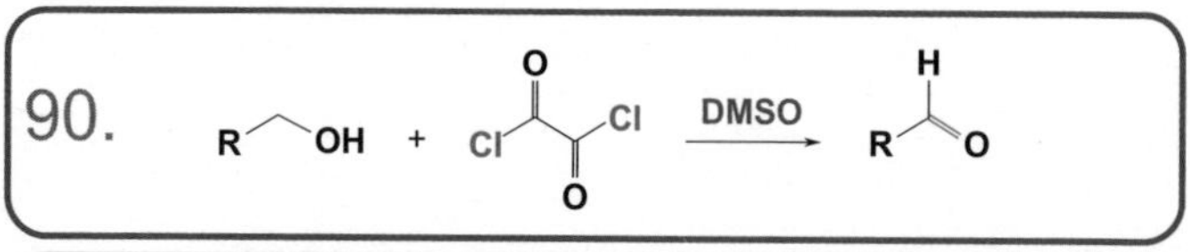

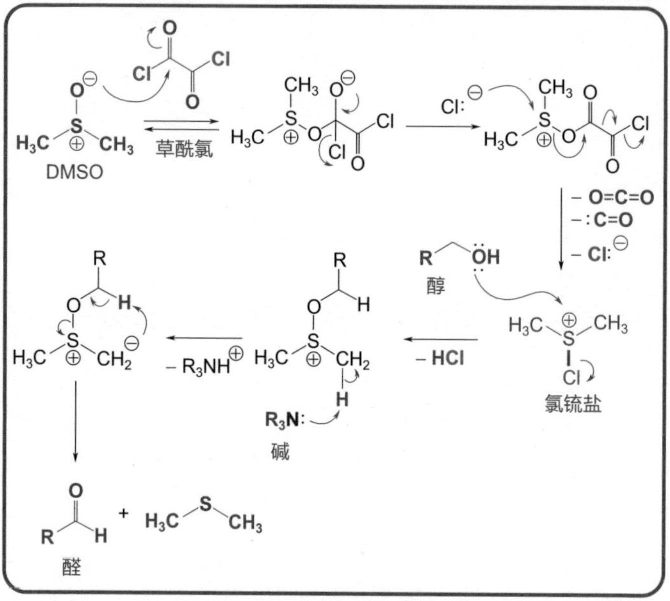

图 90.1　**斯文氧化反应的机理**❶

---

❶[译者注：斯文氧化反应是指以二甲基亚砜（DMSO）为氧化剂，在低温下与草酰氯协同作用将伯醇或仲醇氧化成醛和酮的反应。]斯文氧化是有机合成和药物化学中最重要的反应之一。

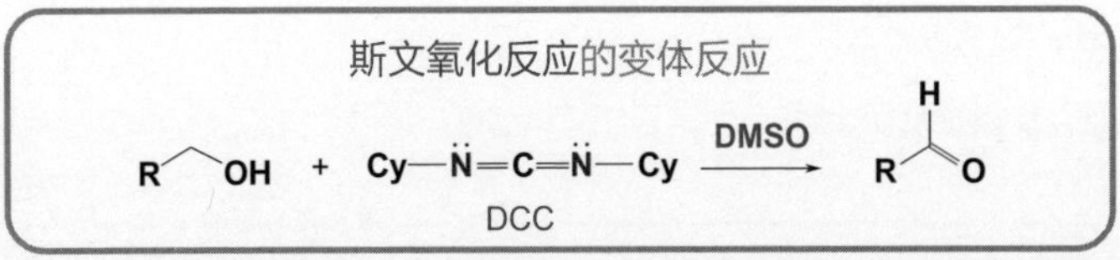

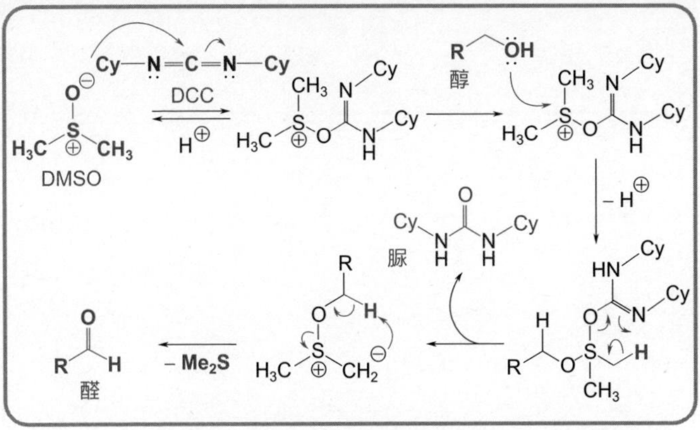

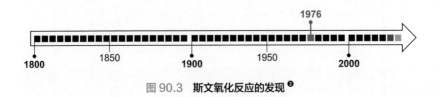

图 90.2 斯文氧化反应的变体反应的机理（DCC + DMSO）❶

图 90.3 斯文氧化反应的发现❷

---

❶ 斯文氧化反应的变体反应有很多，如使用 TFAA 和 DMSO[90a]，或碳二亚胺试剂（DCC）和 DMSO[90b] 的变体反应。几种重要的人名氧化反应可将醇氧化为酮，如戴斯 - 马丁氧化 [Dess-Martin oxidation（参见第 33 个反应机理）] 和琼斯氧化 [Jones oxidation（参见第 51 个反应机理）]。

❷ 该反应大约于 1976 年被首次报道[90a]，详情见参考文献 [90c, 90d]。

91 乌吉反应
(Ugi reaction)

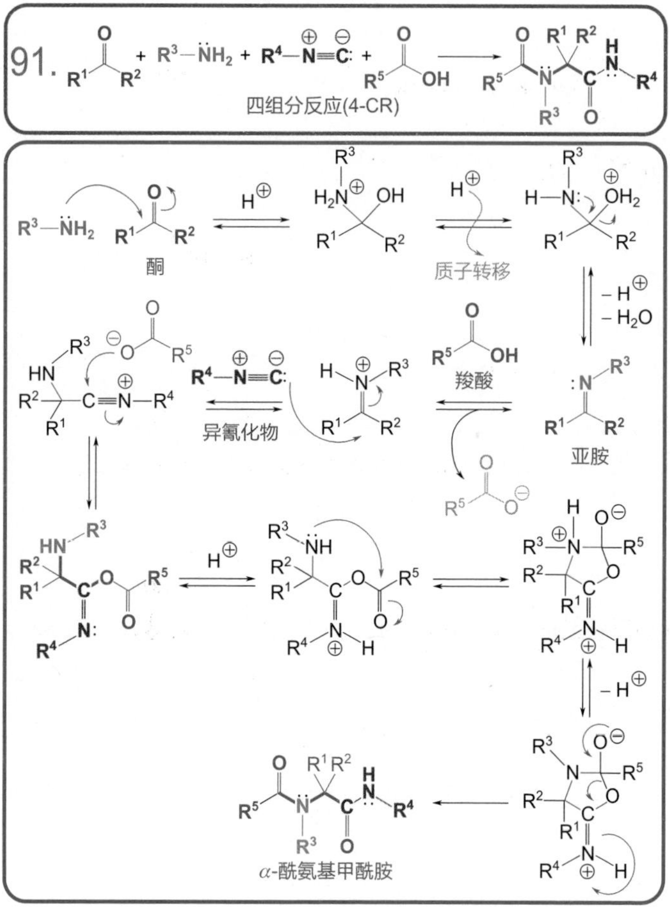

图 91.1 乌吉反应的机理 ❶

---

❶[译者注：乌吉反应也称为乌吉缩合（Ugi condensation），是指一分子醛或酮、一分子胺、一分子异腈以及一分子羧酸缩合生成 α-酰氨基酰胺的反应。]该反应为多组分反应（multi-component reaction，MCR），具体为四组分反应（4-CR）。

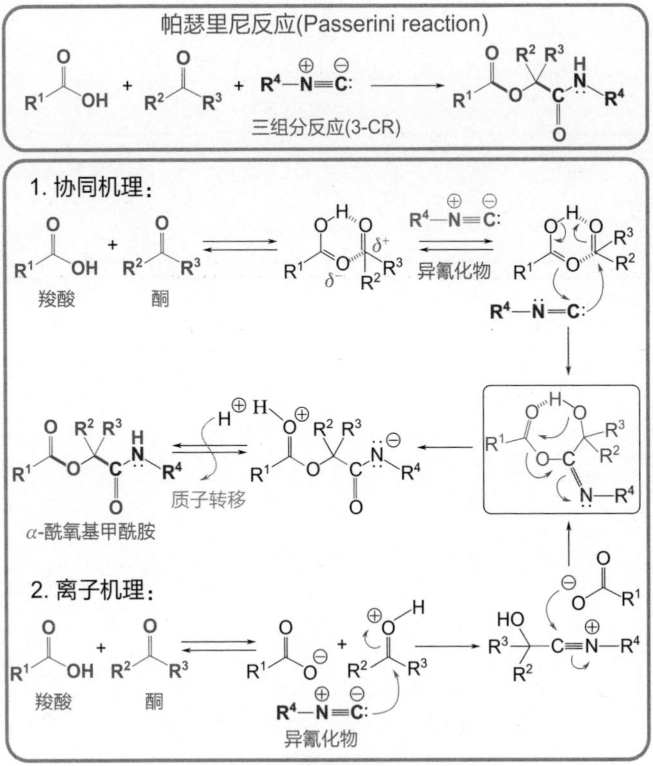

图 91.2 帕瑟里尼反应的机理 ❶

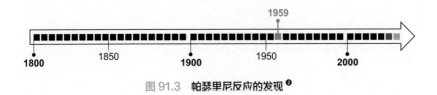

图 91.3 帕瑟里尼反应的发现 ❷

---

❶ 帕瑟里尼反应的机理与乌吉反应相似[91a, 91b]，可以通过协同机理（1）或离子机理（2）解释产物的生成。本书还介绍了其他三组分反应，如格瓦尔德反应（Gewald reaction，第 41 个反应机理）、曼尼希反应（Mannich reaction，第 56 个反应机理）、佩塔西反应（Petasis reaction，第 62 个反应机理）和保森-坎德反应（Pauson-Khand reaction，第 73 个反应机理）。

❷ 该反应大约于 1959 年被首次报道[91c]。

92 乌尔曼芳基-芳基偶联
(Ullmann aryl-aryl coupling)

$$Ar^1-X + Ar^2-X \xrightarrow{[Cu]} Ar^1-Ar^2$$

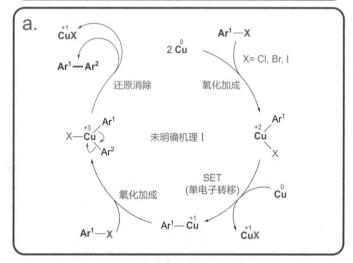

图 92.1　乌尔曼芳基-芳基偶联的机理（a）[1]

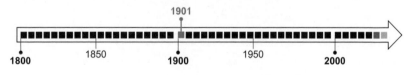

图 92.2　乌尔曼芳基-芳基偶联的发现[2]

---

[1] 乌尔曼芳基-芳基偶联也称为乌尔曼反应（Ullmann reaction），是指卤代芳烃在 Cu、Ni 或 Pd 催化下进行自身偶联生成二芳基化合物的反应。该反应机理尚未完全明确，推测是通过机理 I（a）而形成有机铜中间体 Cu（I）或 Cu（II）。

[2] 该反应大约于 1901 年被首次报道[92a, 92b]。

b.
X
2 Ar
X= Cl, Br, I
2 Cu
SET
2 Cu
X
2 Ar
未明确机理 II
SRN1
Ar Ar
2 Cu
Ar
2 CuX
Ar

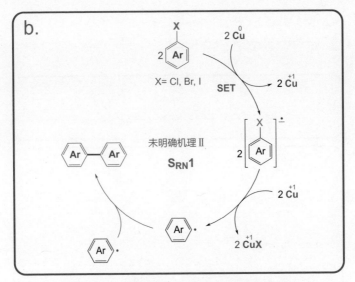

图 92.3　**乌尔曼芳基 – 芳基偶联的机理（b）❶**

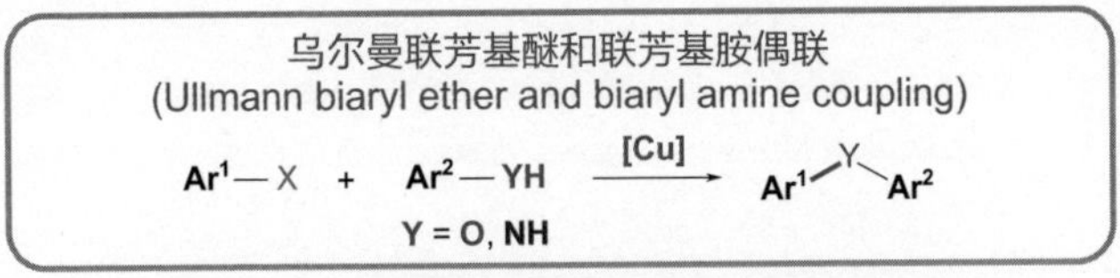

乌尔曼联芳基醚和联芳基胺偶联
(Ullmann biaryl ether and biaryl amine coupling)
Ar1 — X
+
Ar2 — YH
Y = O, NH
[Cu]
Ar1
Y
Ar2

图 92.4　**乌尔曼联芳基醚和联芳基胺偶联反应❷**

---

❶芳香族自由基亲核取代（$S_{RN}1$）机理（参见第 5 个反应机理）形成对称和不对称联芳基产物的另一种解释是机理 II（b）。

❷乌尔曼联芳基醚和联芳基胺的偶联反应在合成上更为实用[92c, 92d]。该反应同样使用芳基卤化物与苯酚或苯胺通过铜介导形成 C—O 和 C—N 键的偶联[92e]。合成芳基醚和胺的另一种方法是通过陈 - 埃文斯 - 兰交叉偶联［Chan-Evans-Lam cross coupling（参见第 23 个反应机理）］。

93 厄普约翰双羟基化反应
(Upjohn dihydroxylation)

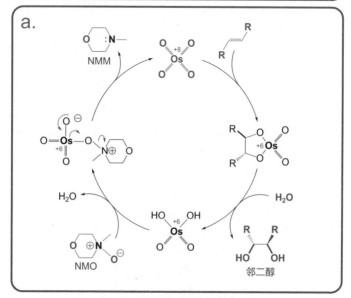

图 93.1 厄普约翰双羟基化反应的机理（a）❶

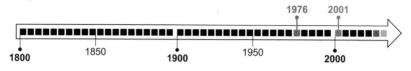

图 93.2 厄普约翰双羟基化反应的发现❷

---

❶[译注：厄普约翰双羟基反应是指以四氧化锇为催化剂、NMO 为氧化剂，将烯烃转化为相应的顺式邻二醇的反应。]机理（a）产生外消旋产物（顺式 -1,2- 乙二醇）[93a]。

❷该反应大约于 1976 年被首次报道[93f]。2001 年，K. Barry Sharpless 和 William S. Knowles、Ryoji Noyori 因手性催化氧化和氢化领域的成就共同获得诺贝尔化学奖[93g]。

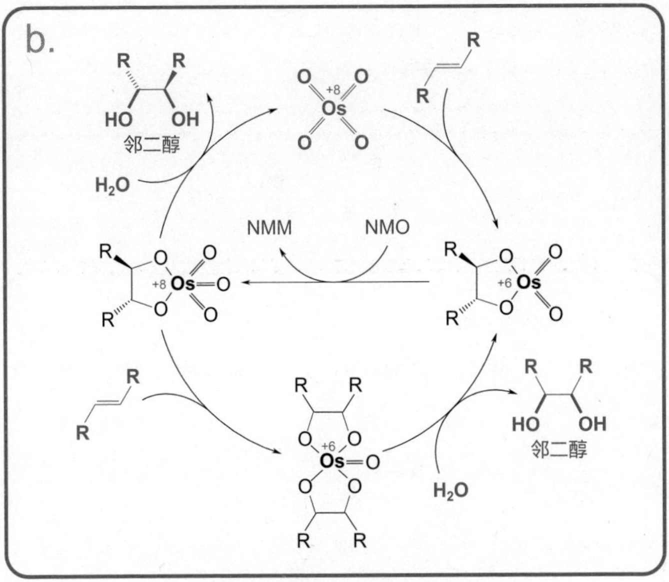

图 93.3 厄普约翰双羟基化反应的机理（b）❶

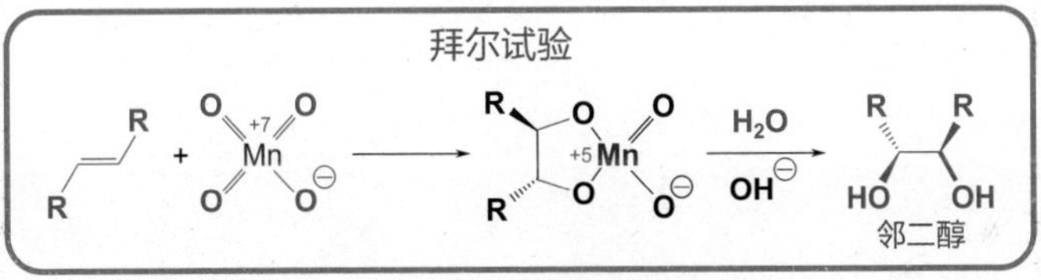

图 93.4 拜尔试验 ❷

---

❶ 沙尔普利斯不对称双羟基化（Sharpless asymmetric dihydroxylation）可以通过机理（b）简单表示。该反应是厄普约翰双羟基化的不对称变体，可产生纯对映体产物[93b, 93c, 93d]。

❷ 拜尔试验（Baeyer test）基于高锰酸钾的 TLC 染色，与厄普约翰双羟基化反应相关联，可检测不饱和双键的存在[93e]。

94 维尔斯迈尔－哈克反应
(Vilsmeier-Haack reaction)

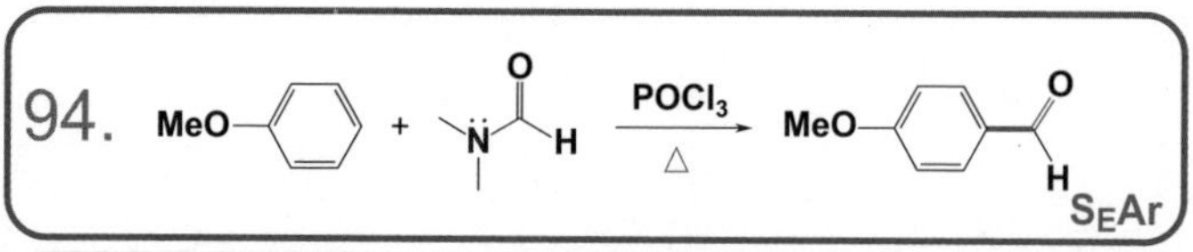

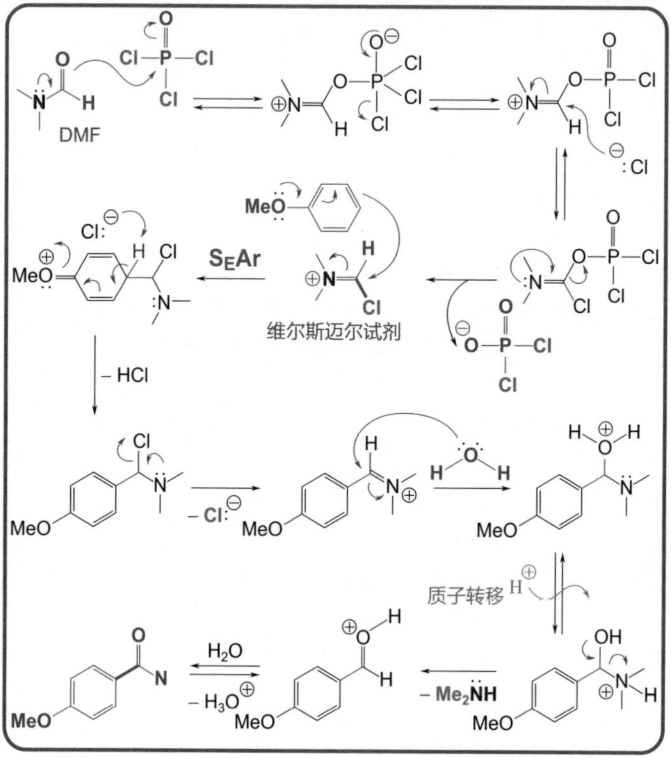

图 94.1 维尔斯迈尔－哈克反应的机理 ❶

---

❶ 维尔斯迈尔-哈克反应，也称为维尔斯迈尔-哈克甲酰化（Vilsmeier-Haack formylation）[译者注：是指芳香化合物与二取代甲酰胺在三氯氧磷作用下，将芳环甲酰化的反应。该反应只适于活泼底物，如苯酚、苯胺]。该反应是芳香族亲电取代的典型实例（芳烃离子机理 =S_EAr，参见第 3 个反应机理）。

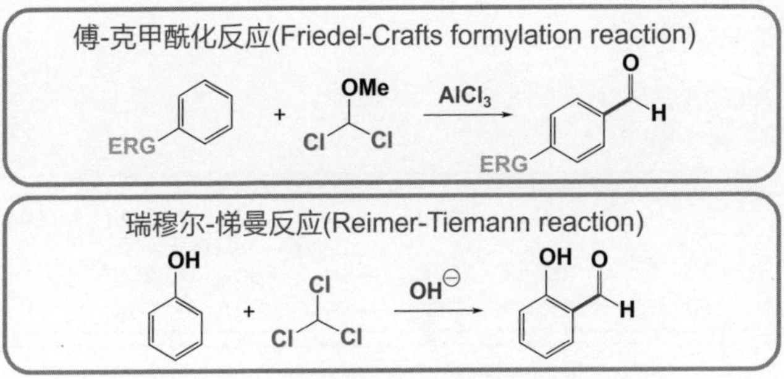

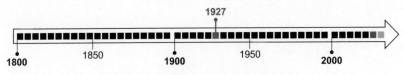

图 94.2　维尔斯迈尔 – 哈克反应的相关反应 ❶

图 94.3　维尔斯迈尔 – 哈克反应的发现 ❷

---

❶ 一些人名反应与维尔斯迈尔 - 哈克反应有关,如使用二氯(甲氧基)甲烷的傅 - 克甲酰化反应(参见第 39 个反应机理)、使用氯仿的瑞穆尔 - 悌曼反应(仅限于酚类的邻位甲酰化反应)[94a]和其他反应[1]。

❷ 该反应大约于 1927 年被首次报道[94b]。

95 瓦克氧化
(Wacker oxidation)

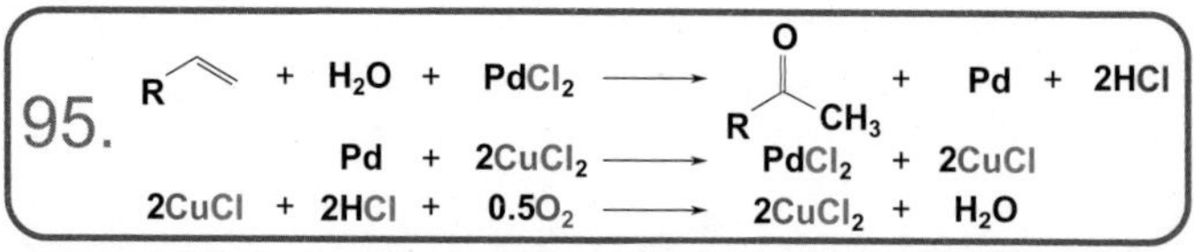

$$\text{R-CH=CH}_2 + H_2O + PdCl_2 \longrightarrow \text{R-CO-CH}_3 + Pd + 2HCl$$
$$Pd + 2CuCl_2 \longrightarrow PdCl_2 + 2CuCl$$
$$2CuCl + 2HCl + 0.5O_2 \longrightarrow 2CuCl_2 + H_2O$$

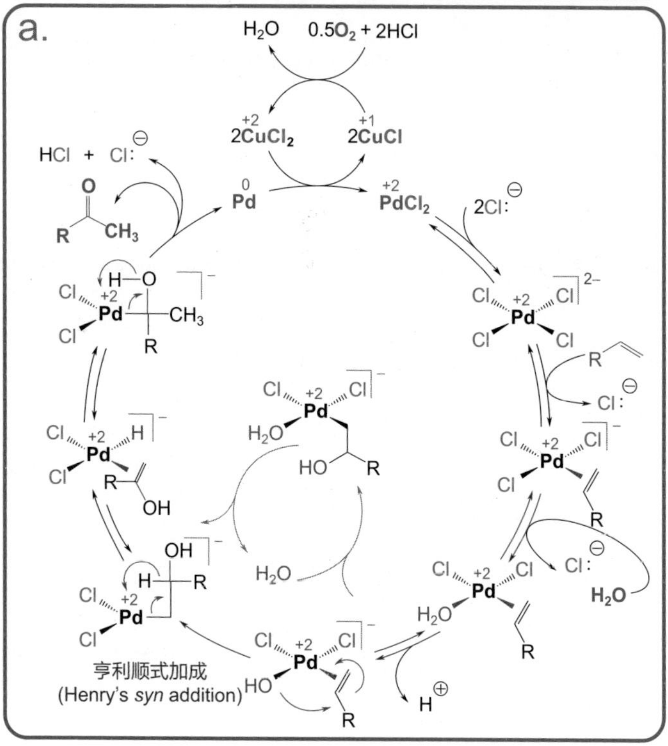

图 95.1　瓦克氧化反应的机理（a）❶

---

❶ 瓦克氧化，也称为瓦克法（Wacker process），是指在氯化钯和氯化铜催化下将烯烃氧化成酮的反应。该反应可能经历不同的反应机理，其中机理（a）亨利顺式加成（内层）是由亨利提出的[95a, 95b]。

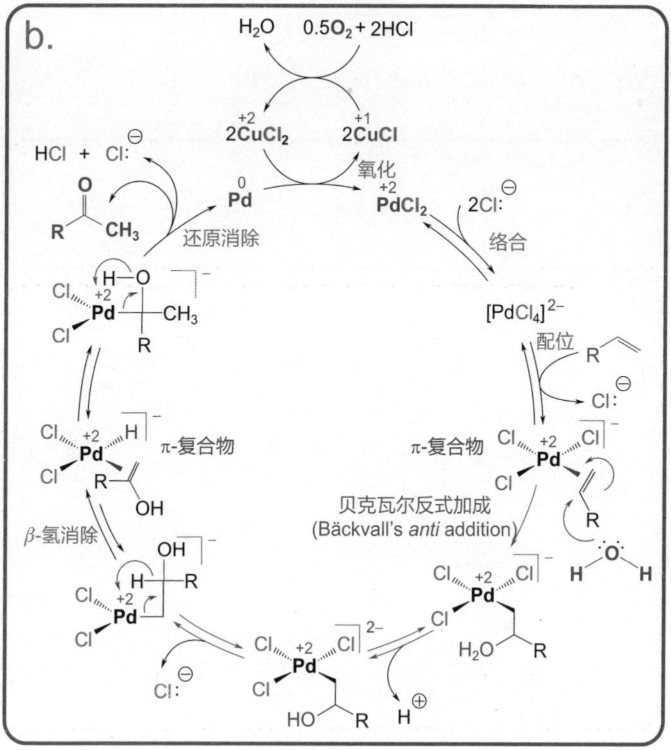

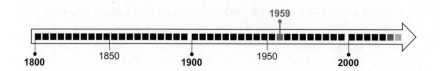

图 95.2 瓦克氧化反应的机理（b）❶

图 95.3 瓦克氧化反应的发现 ❷

---

❶ 贝克瓦尔提出了机理（b），即贝克瓦尔反式加成（外层）[95a, 95b]。
❷ 该反应大约于 1959 年被首次报道[95c]。

96 瓦格纳 – 麦尔外因重排
(Wagner-Meerwein rearrangement)

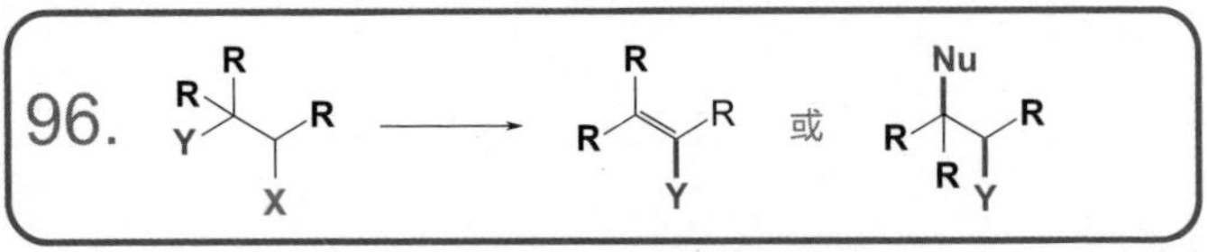

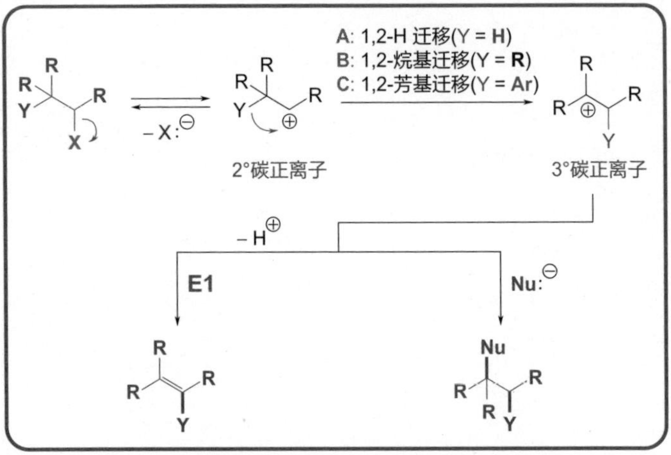

图 96.1 瓦格纳 – 麦尔外因重排反应的一般机理 ❶

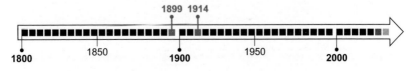

图 96.2 瓦格纳 – 麦尔外因重排反应的发现 ❷

---

❶ 瓦格纳 - 麦尔外因重排是指在酸催化下将醇转换为烯烃的反应,是将新生成的碳正离子重排为更稳定的碳正离子(1° → 2° → 3°)。该反应与频哪醇-频哪酮重排和蒂芬欧-捷姆扬诺夫重排相关(参见第 76 个反应机理)。

❷ 该反应大约于 1899 年和 1914 年分别被瓦格纳[96a, 96b]和麦尔外因[96c]报道。

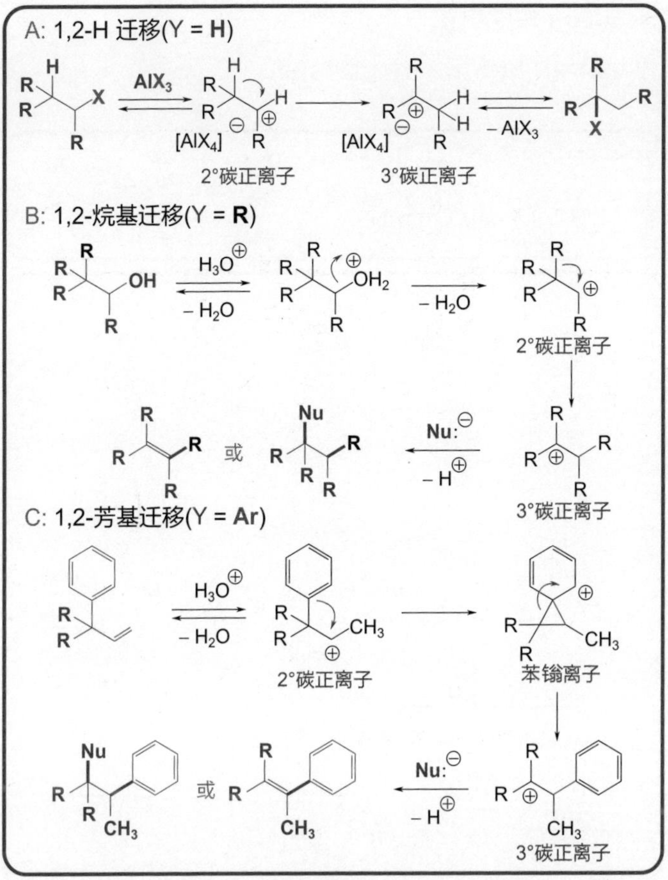

图 96.3　瓦格纳 – 麦尔外因重排反应的机理（A、B、C）❶

---

❶ 该反应生成的碳正离子通过以下三种途径重排生成更稳定的产物：（A）1,2-H 迁移（Y=H）；（B）1,2- 烷基迁移（Y=R）；（C）1,2- 芳基迁移（Y=Ar）。β- 消除反应（E1）通常伴随于瓦格纳 - 麦尔外因重排[1]。

97 温勒伯酮合成
(Weinreb ketone synthesis)

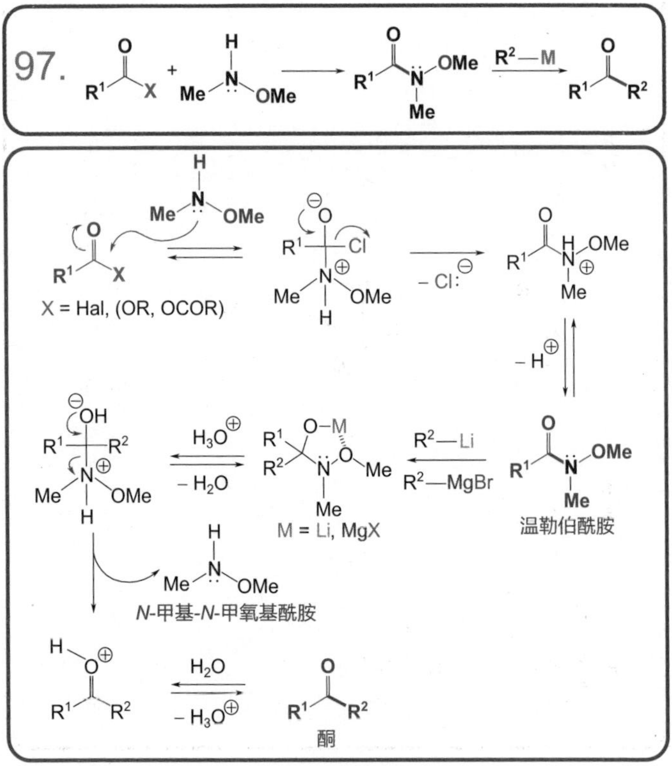

图 97.1　温勒伯酮合成的机理 ❶

---

❶ 温勒伯酮合成是一种借助试剂温勒伯酰胺[温勒伯-纳姆酰胺(Weinreb-Nahm amide)]制备酮的方法[97a]。

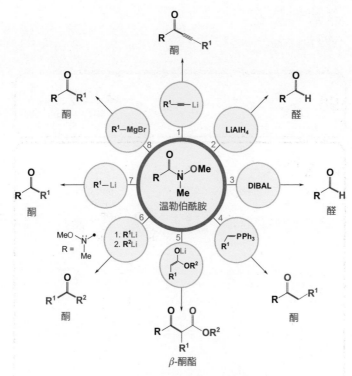

图 97.2　温勒伯酰胺合成的通用性 ❶

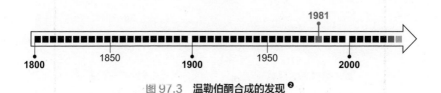

图 97.3　温勒伯酮合成的发现 ❷

---

❶ 温勒伯酰胺具有广泛的合成应用，可与多种亲核试剂反应。这些亲核试剂包括：有机锂和有机镁(格氏试剂)；还原剂，如 DIBAL；磷叶立德或膦烷[97b]、其他试剂等[1]。

❷ 该反应大约于 1981 年被首次报道[97c]。

98 维蒂希反应
(Wittig reaction)

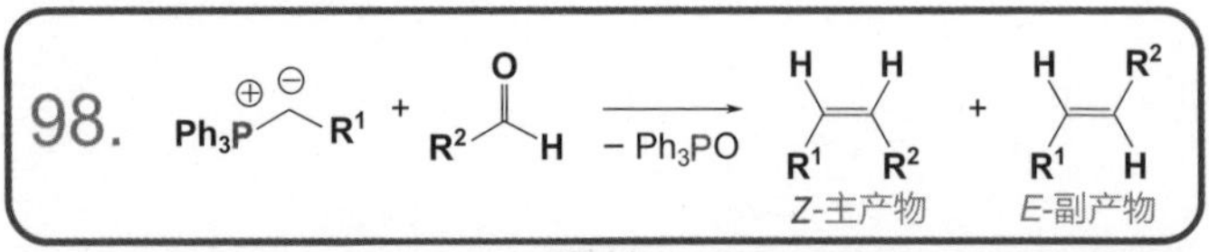

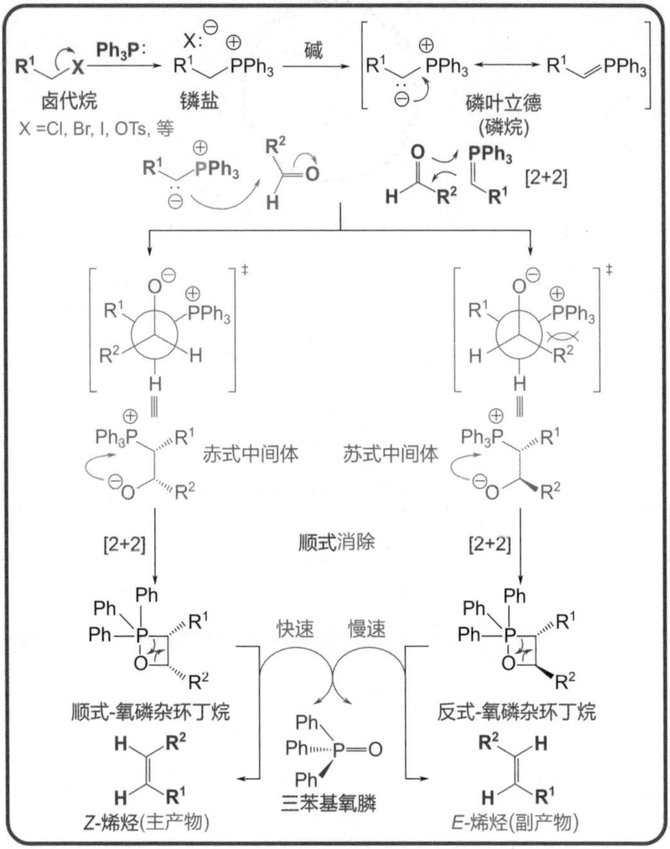

图 98.1 维蒂希反应的机理[1]

---

❶ 维蒂希反应，也称为维蒂希烯烃化（Wittig olefination），是指以磷叶立德（phosphorous ylide）或磷烷（由鏻盐形成）将羰基化合物转化为烯烃的反应[98a]。

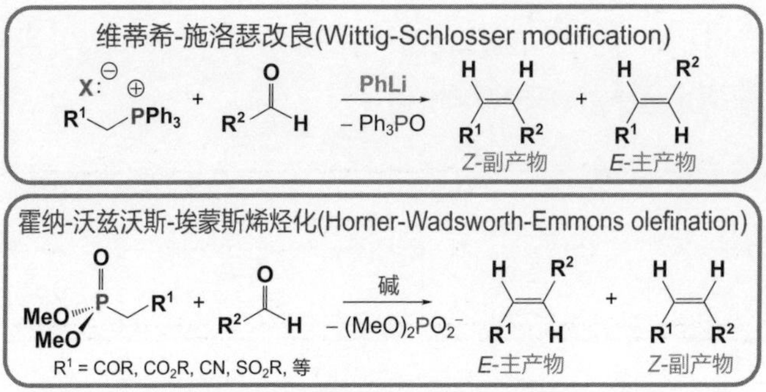

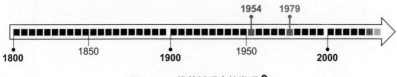

图 98.2　维蒂希反应的相关反应 ❶

图 98.3　维蒂希反应的发现 ❷

---

❶ 维蒂希反应的类似相关反应包括：使用过量碱 PhLi 利于 E-烯烃生成的维蒂希-施洛瑟改良[98b]；依赖膦酸酯 [PO(OR)$_2$R] 的霍纳-沃兹沃斯-埃蒙斯烯烃化（参见第 50 个反应机理），该反应中的膦酸酯可通过阿尔布佐夫反应（第 9 个反应机理）进行制备。

❷ 该反应大约于 1954 年被首次报道[98c, 98d]。1979 年，Georg Wittig 与 Herbert C. Brown 因在磷（和硼）化学领域的贡献而共同获得了诺贝尔化学奖[20c]。

99 沃尔 – 齐格勒反应
(Wohl-Ziegler reaction)

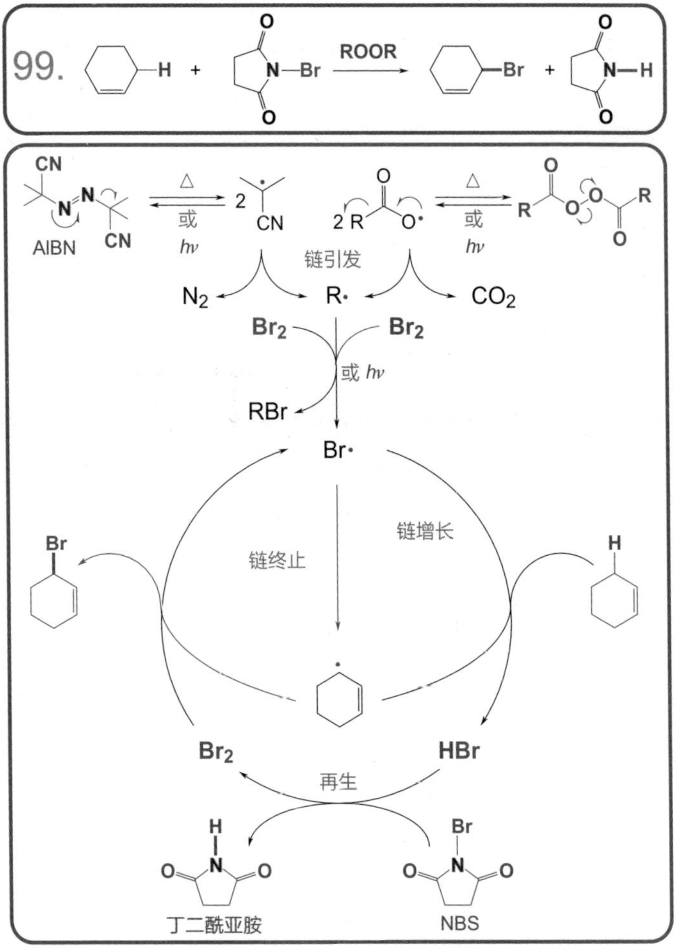

图 99.1 沃尔 – 齐格勒反应的机理 ❶

---

❶ 沃尔 - 齐格勒反应，也称为沃尔 - 齐格勒溴化（Wohl-Ziegler bromination），是一种自由基取代反应，具体是以 N-溴代酰亚胺（NBS）进行的溴化反应（参见第 60 个反应机理中的米尼希反应）。

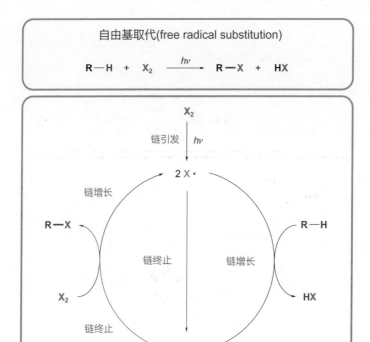

图 99.2　**自由基取代反应的机理** ❶

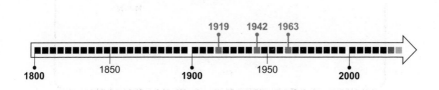

图 99.3　**沃尔－齐格勒反应的发现** ❷

---

❶ 自由基取代机理通常主要有三步：（a）链引发；（b）链增长；（c）链终止。烷烃的自由基氯化反应就是一个典型实例[1]。

❷ 该反应大约于 1919 年和 1942 年分别由 Wohl[99a] 和 Ziegler[99b] 报道。鉴于在该领域的贡献，Karl Ziegler 和 Giulio Natta 于 1963 年共同获得诺贝尔化学奖[99c]。

100 沃尔夫 – 凯惜纳还原
(Wolff-Kishner reduction)

$$100. \quad \underset{R^1}{\overset{O}{\|}}\underset{}{C}R^2 + H_2N-NH_2 \xrightarrow[\Delta]{OH^{\ominus}} \underset{R^1}{\overset{H}{|}}\underset{}{C}\underset{R^2}{\overset{H}{|}} + N_2$$

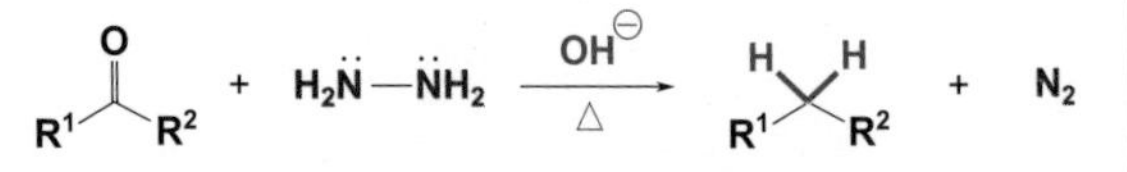

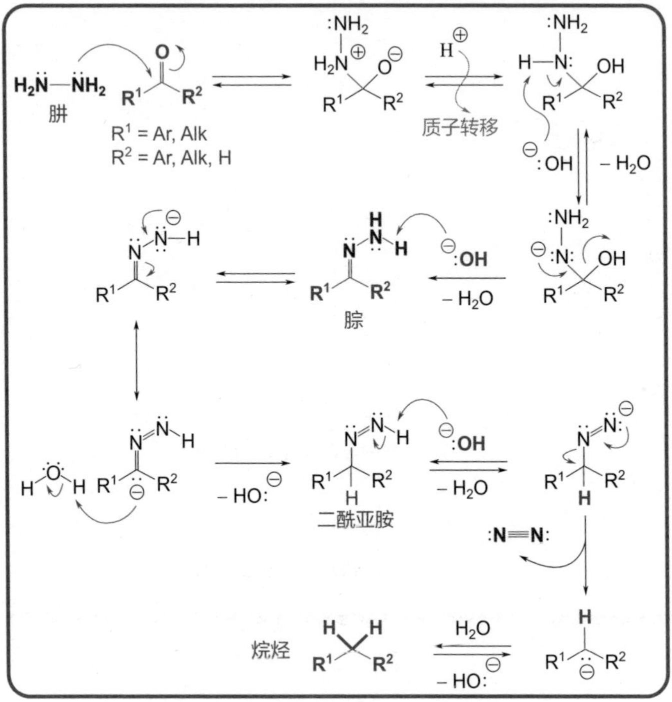

图 100.1　沃尔夫 – 凯惜纳还原反应的机理 ❶

---

❶ [译者注：沃尔夫 - 凯惜纳还原，是指醛或酮的羰基通过形成腙或缩氨脲后在碱性条件下高温分解为亚甲基的反应。] 该反应包括许多改良反应，如黄鸣龙改良（Huang-Minlon modification）等[100a]。

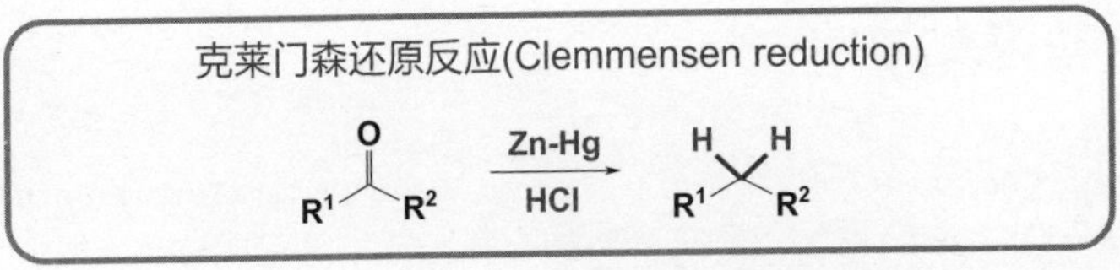

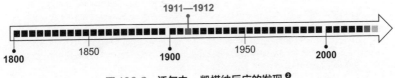

图 100.2　沃尔夫-凯惜纳反应的相关反应 ❶

图 100.3　沃尔夫-凯惜纳反应的发现 ❷

---

❶ 在不考虑反应机理，仅关注生成产物的情况下，克莱门森还原与沃尔夫-凯惜纳还原更为相似[100b]。

❷ 该反应大约最早于 1911 年和 1912 分别由 Kishner[100c] 和 Wolff[100d] 报道。

参考书目与文献

[1] (a) March J. Advanced Organic Chemistry: Reactions, Mechanisms, and Structure (Fourth Edition). New York, NY, USA, J. Wiley & Sons, 1992. (b) Carey FA, and Sundberg RJ. Advanced Organic Chemistry: Part A: Structure and Mechanisms (Third Edition). New York, NY, USA, Plenum Press, 1990. (c) Carey FA, and Sundberg RJ. Advanced Organic Chemistry: Part B: Reactions and Synthesis (Third Edition). New York, NY, USA, Plenum Press, 1990.

[2] (a) Hartwig JF. Organotransition Metal Chemistry: From Bonding to Catalysis. Mill Valley, CA, USA, University Science Books, 2010. (b) Turro NJ, Ramamurthy V, and Scaiano JC. Modern Molecular Photochemistry of Organic Molecules. Sausalito, CA, USA, University Science Books, 2010.

[3] (a) Fleming I. Molecular Orbitals and Organic Chemical Reactions (Reference Edition). Chichester, West Sussex, UK, J. Wiley & Sons, 2010. (b) Fleming I. Molecular Orbitals and Organic Chemical Reactions (Student Edition). Chichester, West Sussex, UK, J. Wiley & Sons, 2009.

[4] (a) Kürti L, and Czakó B. Strategic Applications of Named Reactions in Organic Synthesis: Background and Detailed Mechanisms. Burlington, MA, USA, Elsevier Academic Press, 2005. (b) Li JJ. Name Reactions: A Collection of Detailed Mechanisms and Synthetic Applications (Fifth Edition). Heidelberg, Germany, Springer, 2014.

[5] (a) https://www.organic-chemistry.org/ (accessed December 5, 2019). (b) http://www.name-reaction.com/ (accessedDecember 5, 2019). (c) https://www.synarchive.com/ (accessed December 5, 2019). (d) https://en.wikipedia.org/wiki/Main_Page (accessed December 5, 2019).

[6] e-EROS Encyclopedia of Reagents for Organic Synthesis. Chichester, NY, USA John Wiley & Sons, 2001. The on-line e-EROS database of reagents for organic synthesis can be found at https://onlinelibrary.wiley.com/doi/book/10.1002/047084289X (accessed December 5, 2019).

[7] (a) Bouveault L, and Blanc G. Préparation des alcools primaires au moyen des acides correspondants. *Compt. Rend.* **1903**, 136, 1676–1678. The reference is in French and can be found at https://gallica.bnf.fr/ark:/12148/bpt6k3091c/f1676.image.langFR (accessed December 5, 2019). (b) Bouveault L, and Locquin R. Action du sodium sur les éthers des acides monobasiques à fonction simple de la série grasse. *Compt. Rend.* **1905**, 140, 1593–1595. The reference is in French and can be found at https://gallica.bnf.fr/ark:/12148/bpt6k30949/f1689.image (accessed December 5, 2019).

[8] (a) Faworsky A. Isomerisationserscheinungen der Kohlen-wasserstoffe C_nH_{2n-2}. Erste Abhandlung. *J. Prakt. Chem.* **1888**, 37 (1), 382–395. (b) Faworsky A. Isomerisationserscheinungen der Kohlen-wasserstoffe C_nH_{2n-2}. Zweite Abhandlung. *J. Prakt. Chem.* **1888**, 37 (1), 417–431. (c) Faworsky A. Isomerisationserscheinungen der Kohlen-wasserstoffe C_nH_{2n-2}. Dritte Abhandlung. *J. Prakt. Chem.* **1888**, 37 (1), 531–536. Please note, these references were originally published in Russian and they are difficult to locate. They can be found in German at https://doi.org/10.1002/prac.18880370133; https://doi.org/10.1002/prac.18880370138; and https://doi.org/10.1002/prac.18880370147, respectively (all accessed December 5, 2019). (d) Brown CA, and Yamashita A. Saline hydrides and superbases in organic reactions. IX. Acetylene zipper. Exceptionally facile contrathermodynamic multipositional isomeriazation of alkynes with potassium 3-aminopropylamide. *J. Am. Chem. Soc.* **1975**, 97 (4), 891–892.

[9] (a) See more at https://en.wikipedia.org/wiki/Organophosphorus_compound (accessed December 5, 2019). (b) Michaelis A, and Kaehne R. Ueber das Verhalten der Jodalkyle gegen die sogen. Phosphorigsäureester oder *O*-Phosphine. *Ber. Dtsch. Chem. Ges.* **1898**, 31 (1), 1048–1055. (c) Arbuzov AE. On the structure of phosphonic acid and its derivates: Isometization and transition of bonds from trivalent to pentavalent phosphorus. *J. Russ. Phys. Chem.* Soc. **1906**, 38, 687. Note, the original reference is in Russian and it is difficult to locate. Additional references can be found in [1a].

[10] (a) Wolff L. Ueber Diazoanhydride. *Justus Liebigs Ann. Chem.* **1902**, 325 (2), 129–195. (b) Wolff L, and Krüche R. Über Diazoanhydride (1,2,3-Oxydiazole oder Diazoxyde) und Diazoketone. *Justus Liebigs Ann. Chem.* **1912**, 394 (1), 23–59. (c) Arndt F, and Eistert B. Ein Verfahren zur Überführung von Carbonsäuren in ihre höheren Homologen bzw. deren Derivate. *Ber. Dtsch. Chem. Ges.* A/B **1935**, 68 (1), 200–208.

[11] (a) Dakin HD. The oxidation of hydroxy derivatives of benzaldehyde, acetophenone and related substances. *Am. Chem. J.* **1909**, 42 (6), 477–498. Note, the original reference is difficult to locate. (b) Baeyer A, and Villiger V. Einwirkung des Caro'schen Reagens auf Ketone. *Ber. Dtsch. Chem. Ges.* **1899**, 32 (3), 3625–3633. (c) The Nobel Prize in Chemistry 1905. NobelPrize.org. Nobel Media AB 2019 (accessed December 5, 2019, at https://www.nobelprize.org/prizes/chemistry/1905/summary/).

[12] (a) Barton DHR, Dowlatshahi HA, Motherwell WB, and Villemin D. A new radical decarboxylation reaction for the conversion of carboxylic acids into hydrocarbons. *J. Chem. Soc., Chem. Commun.* **1980**, (15), 732–733. (b) Barton DHR, Crich D, and Motherwell WB. New and improved methods for the radical decarboxylation of acids. *J. Chem. Soc., Chem. Commun.* **1983**, (17), 939–941. (c) Barton DHR, and McCombie SW. A new method for the deoxygenation of secondary alcohols. *J. Chem. Soc., Perkin Trans. 1* **1975**, (16), 1574–1585. (d) Barrett AGM, Prokopiou PA, and Barton DHR. Novel method for the deoxygenation of alcohols. *J. Chem. Soc., Chem. Commun.* **1979**, (24), 1175. (e) The Nobel Prize in Chemistry 1969. NobelPrize.org. Nobel Media AB 2019 (accessed December 5, 2019, at https://www.nobelprize.org/prizes/chemistry/1969/summary/).

[13] (a) Baylis AB, and Hillman MED. German Patent 2155113, **1972**. (b) Morita K, Suzuki Z, and Hirose H. *Bull. Chem. Soc. Jpn.* **1968**, 41 (11), 2815. Note, the original reference can be found at https://www.journal.csj.jp/doi/abs/10.1246/bcsj.41.2815 (accessed December 5, 2019).

[14] Beckmann E. Zur Kenntniss der Isonitrosoverbindungen. *Ber. Dtsch. Chem. Ges.* **1886**, 19 (1), 988–993.

[15] (a) Matsumoto T, Ohishi M, and Inoue S. Selective cross-acyloin condensation catalyzed by thiazolium salt. Formation of 1-hydroxy 2-one from formaldehyde and other aldehydes. *J. Org. Chem.* **1985**, 50 (5), 603–606. (b) Breslow R, and Kool E. A γ-cyclodextrin thiazolium salt holoenzyme mimic for the benzoin condensation. *Tetrahedron Lett.* **1988**, 29 (14), 1635–1638. For the original discovery and additional references please see [1a]. (c) Wöhler, and Liebig. Untersuchungen über das Radikal der Benzoesäure. *Ann. Pharm.* **1832**, 3 (3), 249–282. (d) Lapworth A. XCVI. – Reactions involving the addition of hydrogen cyanide to carbon compounds. *J. Chem. Soc., Trans.* **1903**, 83, 995–1005.

[16] (a) Tadross PM, and Stoltz BM. A Comprehensive History of Arynes in Natural Product Total Synthesis. *Chem. Rev.* **2012**, 112 (6), 3550–3577. (b) Additional references can be found at https://en.wikipedia.org/wiki/Aryne (accessed December 5, 2019) and in [1, 2b]. (c) Roberts JD, Simmons HE, Carlsmith LA, and Vaughan CW. Rearrangement in the reaction of chlorobenzene-1-C[14] with potassium amid. *J. Am. Chem. Soc.* **1953**, 75 (13), 3290–3291. Note, the reaction itself was known for a very long time. It is difficult to locate the original references.

[17] (a) Jones RR, and Bergman RG. p-Benzyne. Generation as an intermediate in a thermal isomerization reaction and trapping evidence for the 1,4-benzenediyl structure. *J. Am. Chem. Soc.* **1972**, 94 (2), 660–661. (b) Bergman RG. Reactive 1,4-dehydroaromatics. *Acc. Chem. Res.* **1973**, 6 (1), 25–31.

[18] (a) Kraus CA. Solutions of Metals in Non-Metallic Solvents; I. General Properties of Solutions of Metals in Liquid Ammonia. *J. Am. Chem. Soc.* **1907**, 29 (11), 1557–1571. (b) Additional references can be found at https://en.wikipedia.org/wiki/Solvated_electron (accessed December 5, 2019). (c) Birch AJ. 117. Reduction by dissolving metals. Part I. *J. Chem. Soc.* **1944**, (0), 430–436.

[19] (a) Pomeranz C. Über eine neue Isochinolinsynthese. *Monatshefte für Chemie* **1893**, 14 (1), 116–119. https://doi.org/10.1007/BF01517862 (accessed December 5, 2019). (b) Fritsch P. Synthesen in der Isocumarin- und Isochinolinreihe. *Ber. Dtsch. Chem. Ges.* **1893**, 26 (1), 419–422. (c) Bischler A, and Napieralski B. Zur Kenntniss einer neuen Isochinolinsynthese. *Ber. Dtsch. Chem. Ges.* **1893**, 26 (2), 1903–1908.

[20] (a) Soderquist JA, Roush WR, and Heo J. (2004). 9-Borabicyclo[3.3.1]nonane Dimer. In e-EROS Encyclopedia of Reagents for Organic Synthesis, (Ed.). doi:10.1002/047084289X.rb235.pub2. (b) Brown HC, and Rao BCS. A new technique for the conversion of olefins into organoboranes and related alcohols. *J. Am. Chem. Soc.* **1956**, 78 (21), 5694–5695. (c) The Nobel Prize in Chemistry 1979. NobelPrize.org. Nobel Media AB 2019 (accessed December 5, 2019, at https://www.nobelprize.org/prizes/chemistry/1979/summary/).

[21] (a) Guram AS, and Buchwald SL. Palladium-Catalyzed Aromatic Aminations with in situ Generated Aminostannanes. *J. Am. Chem. Soc.* **1994**, 116 (17), 7901–7902. (b) Paul F, Patt J, and Hartwig JF. Palladium-catalyzed formation of carbon-nitrogen bonds. Reaction intermediates and catalyst improvements in the hetero cross-coupling of aryl halides and tin amides. *J. Am. Chem. Soc.* **1994**, 116 (13), 5969–5970.

[22] (a) Cannizzaro S. Ueber den der Benzoësäure entsprechenden Alkohol. *Justus Liebigs Ann. Chem.* **1853**, 88 (1), 129–130. (b) For additional references please see: Geissman TA. (2011). The Cannizzaro Reaction. In Organic Reactions, (Ed.). doi:10.1002/0471264180.or002.03.

[23] (a) King AE, Brunold TC, and Stahl SS. Mechanistic Study of Copper-Catalyzed Aerobic Oxidative Coupling of Arylboronic Esters and Methanol: Insights into an Organometallic Oxidase Reaction. *J. Am. Chem. Soc.* **2009**, 131 (14), 5044–5045. (b) King AE, Ryland BL, Brunold TC, and Stahl SS. Kinetic and Spectroscopic Studies of Aerobic Copper(II)-Catalyzed Methoxylation of Arylboronic Esters and Insights into Aryl Transmetalation to Copper(II). *Organometallics* **2012**, 31 (22), 7948–7957. (c) Vantourout JC, Miras HN, Isidro-Llobet A, Sproules S, and Watson AJB. Spectroscopic Studies of the Chan–Lam Amination: A Mechanism-Inspired Solution to Boronic Ester Reactivity. *J. Am. Chem. Soc.* **2017**, 139 (13), 4769–4779. (d) Chan DMT, Monaco KL, Wang RP, and Winter MP. New N- and O-arylations with phenylboronic acids and cupric acetate. *Tetrahedron Lett.* **1998**, 39 (19), 2933–2936. (e) Evans DA, Katz JL, and West TR. Synthesis of diaryl ethers through the copper-promoted arylation of phenols with arylboronic acids. An expedient synthesis of thyroxine. *Tetrahedron Lett.* **1998**, 39 (19), 2937–2942. (f) Lam PYS, Clark CG, Saubern S, Adams J, Winters MP, Chan DMT, and Combs A. New aryl/heteroaryl C-N bond cross-coupling reactions via arylboronic acid/cupric acetate arylation. *Tetrahedron Lett.* **1998**, 39 (19), 2941–2944.

[24] (a) Chichibabin AE, and Zeide OA. New reaction for compounds containing the pyridine nucleus. *J. Russ. Phys. Chem. Soc.* **1914**, 46, 1216–1236. Note, the original reference is published in Russian and it is difficult to locate: Чичибабин А. Е., Зейде О. А. ЖРФХО, 46, 1216 (1914). This reference can be found at https://ru.wikipedia.org/wiki/Реакция_Чичибабина (accessed December 5, 2019). (b) Tschitschibabin AE. Eine neue Darstellungsmethode von Oxyderivaten des Pyridins, Chinolins und ihrer Homologen. *Ber. Dtsch. Chem. Ges.* A/B **1923**, 56 (8), 1879–1885.

[25] (a) Dieckmann W. Ueber ein ringförmiges Analogon des Ketipinsäureesters. *Ber. Dtsch. Chem. Ges.* **1894**, 27 (1), 965–966. (b) Claisen L, and Lowman O. Ueber eine neue Bildungsweise des Benzoylessigäthers. *Ber. Dtsch. Chem. Ges.* **1887**, 20 (1), 651–654.

[26] (a) Additional references and good summary can be found at https://en.wikipedia.org/wiki/Claisen_rearrangement (accessed December 5, 2019). (b) Claisen L. Über Umlagerung von Phenol-allyläthern in C-Allyl-phenole. *Ber. Dtsch. Chem. Ges.* **1912**, 45 (3), 3157–3166.

[27] (a) Sharpless KB, Lauer RF, and Teranishi AY. Electrophilic and nucleophilic organoselenium reagents. New routes to α,β-unsaturated carbonyl compounds. *J. Am. Chem. Soc.* **1973**,

95 (18), 6137–6139. (b) Sharpless KB, Young MW, and Lauer RF. Reactions of selenoxides: Thermal *syn*-elimination and $H_2^{18}O$ exchange. *Tetrahedron Lett.* **1973**, 14 (22), 1979–1982. (c) Cope AC, Foster TT, and Towle PH. Thermal Decomposition of Amine Oxides to Olefins and Dialkylhydroxylamines. *J. Am. Chem. Soc.* **1949**, 71 (12), 3929–3934.

[28] (a) Additional references can be found at https://en.wikipedia.org/wiki/Cope_rearrangement (accessed December 5, 2019). (b) Cope AC, and Hardy EM. The Introduction of Substituted Vinyl Groups. V. A Rearrangement Involving the Migration of an Allyl Group in a Three-Carbon System. *J. Am. Chem. Soc.* **1940**, 62 (2), 441–444.

[29] (a) Criegee R. Eine oxydative Spaltung von Glykolen (II. Mitteil. über Oxydationen mit Blei(IV)-salzen). *Ber. Dtsch. Chem. Ges.* A/B **1931**, 64 (2), 260–266. (b) Malaprade L. Action of polyalcohols on periodic acid. Analytical application. *Bull. Soc. Chim. Fr.* **1928**, 43, 683–696. (c) Malaprade L. A study of the action of polyalcohols on periodic acid and alkaline periodates. *Bull. Soc. Chim. Fr.* **1934**, 5, 833–852. Note, the original references are published in French and they are difficult to locate.

[30] (a) IUPAC. Compendium of Chemical Terminology, 2nd ed. (the "Gold Book"). Compiled by A. D. McNaught and A. Wilkinson. Blackwell Scientific Publications, Oxford (1997). Online version (2019) created by S. J. Chalk. ISBN 0-9678550-9-8. https://doi.org/10.1351/goldbook. The link to the page can be found at https://goldbook.iupac.org/terms/view/C01496 (accessed December 5, 2019). (b) Huisgen R, and Eckell A. 1.3-Dipolare additionen der azomethin-imine. *Tetrahedron Lett.* **1960**, 1 (33), 5–8. (c) Grashey R, Huisgen R, and Leitermann H. 1.3-Dipolare additionen der nitrone. *Tetrahedron Lett.* **1960**, 1 (33), 9–13. (d) Tornøe CW, Christensen C, and Meldal M. Peptidotriazoles on Solid Phase: [1,2,3]-Triazoles by Regiospecific Copper(I)-Catalyzed 1,3-Dipolar Cycloadditions of Terminal Alkynes to Azides. *J. Org. Chem.* **2002**, 67 (9), 3057–3064. (e) Rostovtsev VV, Green LG, Fokin VV, and Sharpless KB. A Stepwise Huisgen Cycloaddition Process: Copper(I)-Catalyzed Regioselective "Ligation" of Azides and Terminal Alkynes. *Angew. Chem. Int. Ed.* **2002**, 41 (14), 2596–2599. (f) Worrell BT, Malik JA, and Fokin VV. Direct Evidence of a Dinuclear Copper Intermediate in Cu(I)-Catalyzed Azide-Alkyne Cycloadditions. *Science* **2013**, 340 (6131), 457–460.

[31] (a) Curtius T. Ueber Stickstoffwasserstoffsäure (Azoimid) N_3H. *Ber. Dtsch. Chem. Ges.* **1890**, 23 (2), 3023–3033. (b) Curtius T. 20. Hydrazide und Azide organischer Säuren I. Abhandlung. *J. Prakt. Chem.* **1894**, 50 (1), 275–294. (c) Schmidt KF. Aus den Sitzungen der Abteilungen. *Angew. Chem.* **1923**, 36 (57), 506–523. The reference can be found at https://doi.org/10.1002/ange.19230366703 (accessed December 5, 2019). (d) Schmidt KF. Über den Imin-Rest. *Ber. Dtsch. Chem. Ges.* A/B **1924**, 57, 704–706. (e) Hofmann AW. Ueber die Einwirkung des Broms in alkalischer Lösung auf Amide. *Ber. Dtsch. Chem. Ges.* **1881**, 14 (2), 2725–2736. (f) Lossen W. Ueber Benzoylderivate des Hydroxylamins. *Justus Liebigs Ann. Chem.* **1872**, 161 (2–3), 347–362.

[32] (a) In 1990 Elias James Corey received the Nobel Prize in Chemistry: The Nobel Prize in Chemistry 1990. NobelPrize.org. Nobel Media AB 2019 (accessed December 5, 2019, at https://www.nobelprize.org/prizes/chemistry/1990/summary/). (b) Corey EJ, and Chaykovsky M. Dimethyloxosulfonium Methylide ((CH_3)$_2$SOCH$_2$) and Dimethylsulfonium Methylide ((CH_3)$_2$SCH$_2$). Formation and Application to Organic Synthesis. *J. Am. Chem. Soc.* **1965**, 87 (6), 1353–1364. (c) Darzens G. Method generale de synthese des aldehyde a l'aide des acides glycidique substitues. *Compt. Rend.* **1904**, 139, 1214–1217. The reference is in French and can be accessed at https://gallica.bnf.fr/ark:/12148/bpt6k30930/f1214.image.langFR (accessed December 5, 2019).

[33] (a) Boeckman RJ, and George KM. (2009). 1,1,1-Triacetoxy-1,1-dihydro-1,2-benziodoxol-3(1*H*)-one.In e-EROS Encyclopedia of Reagents for Organic Synthesis, (Ed.). doi:10.1002/047084289X.rt157m.pub2. (b) Zhdankin VV. Organoiodine(V) Reagents in Organic

Synthesis. *J. Org. Chem.* **2011**, 76 (5), 1185–1197. (c) Dess DB, and Martin JC. Readily accessible 12-I-5 oxidant for the conversion of primary and secondary alcohols to aldehydes and ketones. *J. Org. Chem.* **1983**, 48 (22), 4155–4156.

[34] (a) This term appears in some publications and can be found at https://en.wikipedia.org/wiki/Diazonium_compound (accessed December 5, 2019). (b) Griess P. Vorläufige Notiz über die Einwirkung von salpetriger Säure auf Amidinitro- und Aminitrophenylsäure. *Annalen der Chemie und Pharmacie* **1858**, 106 (1), 123–125. The original reference is in German and can be found at https://babel.hathitrust.org/cgi/pt?id=njp.32101044011037&view=1up&seq=541 (accessed December 5, 2019).

[35] (a) Diels O, and Alder K. Synthesen in der hydroaromatischen Reihe. *Justus Liebigs Ann. Chem.* **1928**, 460 (1), 98–122. (b) Additional summary of 27 references can be found at https://en.wikipedia.org/wiki/Diels%E2%80%93Alder_reaction (accessed December 5, 2019). (c) The Nobel Prize in Chemistry 1950. NobelPrize.org. Nobel Media AB 2019 (accessed December 5, 2019, at https://www.nobelprize.org/prizes/chemistry/1950/summary/).

[36] (a) Zimmerman HE, and Grunewald GL. The Chemistry of Barrelene. III. A Unique Photoisomerization to Semibullvalene. *J. Am. Chem. Soc.* **1966**, 88 (1), 183–184. (b) Zimmerman HE, Binkley RW, Givens RS, and Sherwin MA. Mechanistic organic photochemistry. XXIV. The mechanism of the conversion of barrelene to semibullvalene. A general photochemical process. *J. Am. Chem. Soc.* **1967**, 89 (15), 3932–3933.

[37] (a) Smissman EE, and Hite G. The Quasi-Favorskii Rearrangement. I. The Preparation of Demerol and β-Pethidine. *J. Am. Chem. Soc.* **1959**, 81 (5), 1201–1203. (b) Smissman EE, and Hite G. The Quasi-Favorskii Rearrangement. II. Stereochemistry and Mechanism. *J. Am. Chem. Soc.* **1960**, 82 (13), 3375–3381. (c) Favorskii AE. *J. Russ. Phys. Chem. Soc.* **1894**, 26, 590. Note, the original reference is published in Russian and it is difficult to locate: Favorskii AE. *Zh. Russ. Khim. Obshch.* **1894**, 26, 590. (d) Faworsky A. Über die Einwirkung von Phosphorhalogenverbindungen auf Ketone, Bromketone und Ketonalkohole. *J. Prakt. Chem.* **1913**, 88 (1), 641–698.

[38] (a) Shine HJ, Zmuda H, Park KH, Kwart H, Horgan AG, Collins C, and Maxwell BE. Mechanism of the benzidine rearrangement. Kinetic isotope effects and transition states. Evidence for concerted rearrangement. *J. Am. Chem. Soc.* **1981**, 103 (4), 955–956. (b) Fischer E, and Jourdan F. Ueber die Hydrazine der Brenztraubensäure. *Ber. Dtsch. Chem. Ges.* **1883**, 16 (2), 2241–2245. (c) Fischer E, and Hess O. Synthese von Indolderivaten. *Ber. Dtsch. Chem. Ges.* **1884**, 17 (1), 559–568. (d) The Nobel Prize in Chemistry 1902. NobelPrize.org. Nobel Media AB 2019 (accessed December 5, 2019, at https://www.nobelprize.org/prizes/chemistry/1902/summary/).

[39] (a) Ador E, and Crafts J. Ueber die Einwirkung des Chlorkohlenoxyds auf Toluol in Gegenwart von Chloraluminium. *Ber. Dtsch. Chem. Ges.* **1877**, 10 (2), 2173–2176. (b) Friedel C, and Crafts JM. A New General Synthetical Method of Producing Hydrocarbons. *J. Chem. Soc.* **1877**, 32 (0), 725–791. The reference can be found at DOI: 10.1039/JS8773200725 (accessed December 5, 2019).

[40] (a) Ing HR, and Manske RHF. CCCXII. – A modification of the Gabriel synthesis of amines. *J. Chem. Soc.* **1926**, 129 (0), 2348–2351. (b) Delépine M. Sur l'hexamethylene-amine (suite). Solubilities, hydrate, bromure, sulfate, phosphate. *Bull. Soc. Chim. Fr.* **1895**, 13 (3), 352–361. Note, the original reference is published in French and it is difficult to locate. (c) Gabriel S. Ueber eine Darstellungsweise primärer Amine aus den entsprechenden Halogenverbindungen. *Ber. Dtsch. Chem. Ges.* **1887**, 20 (2), 2224–2236.

[41] (a) Knoevenagel E. Condensation von Malonsäure mit aromatischen Aldehyden durch Ammoniak und Amine. *Ber. Dtsch. Chem. Ges.* **1898**, 31 (3), 2596–2619. (b) Gewald K, Schinke E, and Böttcher H. Heterocyclen aus CH-aciden Nitrilen, VIII. 2-Amino-thiophene aus methylenaktiven Nitrilen, Carbonylverbindungen und Schwefel. *Chem. Ber.* **1966**, 99 (1), 94–100.

[42] (a) Eglinton G, and Galbraith AR. 182. Macrocyclic acetylenic compounds. Part I. *Cyclo*tetradeca-1:3-diyne and related compounds. *J. Chem. Soc.* **1959**, (0), 889–896. (b) Behr OM, Eglinton G, Galbraith AR, and Raphael RA. 722. Macrocyclic acetylenic compounds. Part II. 1,2:7,8-Dibenzocyclododeca-1,7-diene-3,5,9,11-tetrayne. *J. Chem. Soc.* **1960**, (0), 3614–3625. (c) Glaser C. Beiträge zur Kenntniss des Acetenylbenzols. *Ber. Dtsch. Chem. Ges.* **1869**, 2 (1), 422–424. (d) Hay A. Communications- Oxidative Coupling of Acetylenes. *J. Org. Chem.* **1960**, 25 (7), 1275–1276. (e) Hay AS. Oxidative Coupling of Acetylenes. II *J. Org. Chem.* **1962**, 27 (9), 3320–3321. (f) Shi W, and Lei A. 1,3-Diyne chemistry: synthesis and derivations. *Tetrahedron Lett.* **2014**, 55 (17), 2763–2772. See also references therein.

[43] (a) Grignard V. Sur quelques nouvelles combinaisons organométaliques du magnésium et leur application à des synthèses d'alcools et d'hydrocabures. *Compt. Rend.* **1900**, 130, 1322–1324. The original reference is in French and can be accessed at https://gallica.bnf.fr/ark:/12148/bpt6k3086n/f1322.table (accessed December 5, 2019). (b) The Nobel Prize in Chemistry 1912. NobelPrize.org. Nobel Media AB 2019 (accessed December 5, 2019, at https://www.nobelprize.org/prizes/chemistry/1912/summary/).

[44] (a) Prantz K, and Mulzer J. Synthetic Applications of the Carbonyl Generating Grob Fragmentation. *Chem. Rev.* **2010**, 110 (6), 3741–3766. (b) Grob CA, and Baumann W. Die 1,4-Eliminierung unter Fragmentierung. *Helv. Chim. Acta* **1955**, 38 (3), 594–610. (c) Grob CA, and Schiess PW. Heterolytic Fragmentation. A Class of Organic Reactions. *Angew. Chem. Int. Ed. Engl.* **1967**, 6 (1), 1–15.

[45] (a) Surellas GS. Notes sur l'Hydriodate de potasse et l'Acide hydriodique. – Hydriodure de carbone; moyen d'obtenir, à l'instant, ce composé triple. **1822**. The original reference is in French and can be accessed at https://gallica.bnf.fr/ark:/12148/bpt6k6137757n/f2.image (accessed December 5, 2019). (b) Liebig J. Ueber die Verbindungen, welche durch die Einwirkung des Chlors auf Alkohol, Aether, ölbildendes Gas und Essiggeist entstehen. *Ann. Phys.* **1832**, 100 (2), 243–295. (c) Lieben A. Ueber Entstehung von Jodoform und Anwendung dieser Reaction in der chemischen Analyse. *Annalen der Chemie. Supplementband.* **1870**, 7, 218–236. The original reference is in German and can be accessed at https://babel.hathitrust.org/cgi/pt?id=uiug.30112018225695&view=1up&seq=230 (accessed December 5, 2019). Note, additional references can be found at https://en.wikipedia.org/wiki/Haloform_reaction (accessed December 5, 2019).

[46] (a) Heck RF. Acylation, methylation, and carboxyalkylation of olefins by Group VIII metal derivatives. *J. Am. Chem. Soc.* **1968**, 90 (20), 5518–5526. (b) Mizoroki T, Mori K, and Ozaki A. Arylation of Olefin with Aryl Iodide Catalyzed by Palladium. *Bull. Chem. Soc. Jpn.* **1971**, 44 (2), 581. (c) The Nobel Prize in Chemistry 2010. NobelPrize.org. Nobel Media AB 2019 (accessed December 5, 2019, at https://www.nobelprize.org/prizes/chemistry/2010/summary/).

[47] (a) Hell C. Ueber eine neue Bromirungsmethode organischer Säuren. *Ber. Dtsch. Chem. Ges.* **1881**, 14 (1), 891–893. (b) Volhard J. 4) Ueber Darstellung α-bromirter Säuren. *Justus Liebigs Ann. Chem.* **1887**, 242 (1–2), 141–163. (c) Zelinsky N, Ueber eine bequeme Darstellungsweise von α-Brompropionsäureester. *Ber. Dtsch. Chem. Ges.* **1887**, 20 (1), 2026.

[48] (a) Denmark SE, and Regens CS. Palladium-Catalyzed Cross-Coupling Reactions of Organosilanols and Their Salts: Practical Alternatives to Boron- and Tin-Based Methods. *Acc. Chem. Res.* **2008**, 41 (11), 1486–1499. (b) Hatanaka Y, and Hiyama T. Cross-coupling of organosilanes with organic halides mediated by a palladium catalyst and tris(diethylamino)sulfonium difluorotrimethylsilicate. *J. Org. Chem.* **1988**, 53 (4), 918–920.

[49] (a) Saytzeff A. Zur Kenntniss der Reihenfolge der Analgerung und Ausscheidung der Jodwasserstoffelemente in organischen Verbindungen. *Justus Liebigs Ann. Chem.* **1875**, 179 (3), 296–301. (b) Cope AC, and Trumbull ER. (2011). Olefins from Amines: The Hofmann Elimination Reaction and Amine Oxide Pyrolysis. In Organic Reactions, (Ed.).

doi:10.1002/0471264180.or011.05. (c) Hofmann AW. Beiträge zur Kenntniss der flüchtigen organischen Basen. *Justus Liebigs Ann. Chem.* **1851**, 78 (3), 253–286. (d) Hofmann AW. Beiträge zur Kenntniss der flüchtigen organischen Basen. *Justus Liebigs Ann. Chem.* **1851**, 79 (1), 11–39.

[50] (a) Maryanoff BE, and Reitz AB. The Wittig olefination reaction and modifications involving phosphoryl-stabilized carbanions. Stereochemistry, mechanism, and selected synthetic aspects. *Chem. Rev.* **1989**, 89 (4), 863–927. (b) Peterson DJ. Carbonyl olefination reaction using silyl-substituted organometallic compounds. *J. Org. Chem.* **1968**, 33 (2), 780–784. (c) Horner L, Hoffmann H, and Wippel HG. Phosphororganische Verbindungen, XII. Phosphinoxyde als Olefinierungsreagenzien. *Chem. Ber.* **1958**, 91 (1), 61–63. (d) Horner L, Hoffmann H, Wippel HG, and Klahre G. Phosphororganische Verbindungen, XX. Phosphinoxyde als Olefinierungsreagenzien. *Chem. Ber.* **1959**, 92 (10), 2499–2505. (e) Wadsworth WS, and Emmons WD. The Utility of Phosphonate Carbanions in Olefin Synthesis. *J. Am. Chem. Soc.* **1961**, 83 (7), 1733–1738.

[51] (a) Freeman F. (2001). Chromic Acid. In e-EROS Encyclopedia of Reagents for Organic Synthesis, (Ed.). doi:10.1002/047084289X.rc164. (b) Piancatelli G, and Luzzio FA. (2007). Pyridinium Chlorochromate. In e-EROS Encyclopedia of Reagents for Organic Synthesis, (Ed.). doi:10.1002/9780470842898.rp288.pub2. (c) Piancatelli G. (2001). Pyridinium Dichromate. In e-EROS Encyclopedia of Reagents for Organic Synthesis, (Ed.). doi:10.1002/047084289X.rp290. (d) Bowden K, Heilbron IM, Jones ERH, and Weedon BCL. 13. Researches on acetylenic compounds. Part I. The preparation of acetylenic ketones by oxidation of acetylenic carbinols and glycols. *J. Chem. Soc.* **1946**, (0), 39–45.

[52] (a) Markownikoff W. I. Ueber die Abhängigkeit der verschiedenen Vertretbarkeit des Radicalwasserstoffs in den isomeren Buttersäuren. *Justus Liebigs Ann. Chem.* **1870**, 153 (2), 228–259. (b) Kutscheroff M. Ueber eine neue Methode direkter Addition von Wasser (Hydratation) an die Kohlenwasserstoffe der Acetylenreihe. *Ber. Dtsch. Chem. Ges.* **1881**, 14 (1), 1540–1542.

[53] (a) Corriu RJP, and Masse JP. Activation of Grignard reagents by transition-metal complexes. A new and simple synthesis of *trans*-stilbenes and polyphenyls. *J. Chem. Soc., Chem. Commun.* **1972**, (3), 144a. (b) Tamao K, Sumitani K, and Kumada M. Selective carbon-carbon bond formation by cross-coupling of Grignard reagents with organic halides. Catalysis by nickel-phosphine complexes. *J. Am. Chem. Soc.* **1972**, 94 (12), 4374–4376. (c) Tamao K, Kiso Y, Sumitani K, and Kumada M. Alkyl group isomerization in the cross-coupling reaction of secondary alkyl Grignard reagents with organic halides in the presence of nickel-phosphine complexes as catalysts. *J. Am. Chem. Soc.* **1972**, 94 (26), 9268–9269.

[54] (a) Ley SV, Norman J, and Wilson AJ. (2011). Tetra-*n*-propylammonium Perruthenate. In e-EROS Encyclopedia of Reagents for Organic Synthesis, (Ed.). doi:10.1002/047084289X.rt074.pub2. (b) Griffith WP, Ley SV, Whitcombe GP, and White AD. Preparation and use of tetra-n-butylammonium per-ruthenate (TBAP reagent) and tetra-n-propylammonium per-ruthenate (TPAP reagent) as new catalytic oxidants for alcohols. *J. Chem. Soc., Chem. Commun.* **1987**, (21), 1625–1627.

[55] (a) Liebeskind LS, and Srogl J. Thiol Ester–Boronic Acid Coupling. A Mechanistically Unprecedented and General Ketone Synthesis. *J. Am. Chem. Soc.* **2000**, 122 (45), 11260–11261. (b) Cheng HG, Chen H, Liu Y, and Zhou Q. The Liebeskind–Srogl Cross-Coupling Reaction and its Synthetic Applications. *Asian J. Org. Chem.* **2018**, 7 (3), 490–508. The open access paper can be found at https://onlinelibrary.wiley.com/doi/epdf/10.1002/ajoc.201700651 (accessed December 5, 2019).

[56] (a) Kleinman EF. (2001). Dimethyl(methylene)ammonium Iodide. In e-EROS Encyclopedia of Reagents for Organic Synthesis, (Ed.). doi:10.1002/047084289X.rd346. (b) Mannich C, and Krösche W. Ueber ein Kondensationsprodukt aus Formaldehyd, Ammoniak und Antipyrin. *Arch. Pharm. Pharm. Med. Chem.* **1912**, 250 (1), 647–667.

[57] (a) Fittig R. Ueber einige Producte der trockenen Destillation essigsaurer Salze. *Justus Liebigs Ann. Chem.* **1859**, 110 (1), 17–23. (b) Demselben. Ueber einige Metamorphosen des Acetons der Essigsäure. *Justus Liebigs Ann. Chem.* **1859**, 110 (1), 23–45. (c) McMurry JE, and Fleming MP. New method for the reductive coupling of carbonyls to olefins. Synthesis of β-carotene. *J. Am. Chem. Soc.* **1974**, 96 (14), 4708–4709.

[58] (a) Meerwein H, and Schmidt R. Ein neues Verfahren zur Reduktion von Aldehyden und Ketonen. *Justus Liebigs Ann. Chem.* **1925**, 444 (1), 221–238. (b) Verley A. Sur l'échange de groupements fonctionnels entre deux molécules. Passage de la fonction alcool à la fonction aldéhyde et inversement. *Bull. Soc. Chim. Fr.* **1925**, 37, 537–542. Note, the original reference is published in French and it is difficult to locate. (c) Ponndorf W. Der reversible Austausch der Oxydationsstufen zwischen Aldehyden oder Ketonen einerseits und primären oder sekundären Alkoholen anderseits. *Angew. Chem.* **1926**, 39 (5), 138–143.

[59] (a) Stetter H, and Schreckenberg M. A New Method for Addition of Aldehydes to Activated Double Bonds. *Angew. Chem. Int. Ed. Engl.* **1973**, 12 (1), 81. (b) Michael A. Ueber die Addition von Natriumacetessig- und Natriummalonsäureäthern zu den Aethern ungesättigter Säuren. *J. Prakt. Chem.* **1887**, 35 (1), 349–356.

[60] (a) Hong Y. (2001). Hydrogen Peroxide–Iron(II) Sulfate. In e-EROS Encyclopedia of Reagents for Organic Synthesis, (Ed.). doi:10.1002/047084289X.rh043. (b) Mihailović ML, Čeković Ž, and Mathes BM. (2005). Lead(IV) Acetate. In e-EROS Encyclopedia of Reagents for Organic Synthesis, (Ed.). doi:10.1002/047084289X.rl006.pub2. (c) Kolbe H. Untersuchungen über die Elektrolyse organischer Verbindungen. *Justus Liebigs Ann. Chem.* **1849**, 69 (3), 257–294. (d) Minisci F, Galli R, Cecere M, Malatesta V, and Caronna T. Nucleophilic character of alkyl radicals: new syntheses by alkyl radicals generated in redox processes. *Tetrahedron Lett.* **1968**, 9 (54), 5609–5612. (e) Minisci F, Bernardi R, Bertini F, Galli R, and Perchinummo M. Nucleophilic character of alkyl radicals – VI: A new convenient selective alkylation of heteroaromatic bases. *Tetrahedron* **1971**, 27 (15), 3575–3579.

[61] (a) Jenkins ID, and Mitsunobu O. (2001). Triphenylphosphine–Diethyl Azodicarboxylate. In e-EROS Encyclopedia of Reagents for Organic Synthesis, (Ed.). doi:10.1002/047084289X.rt372. (b) Hughes DL. (2004). The Mitsunobu Reaction. In Organic Reactions, (Ed.). doi:10.1002/0471264180.or042.02. (c) Swamy KCK, Kumar NNB, Balaraman E, and Kumar KVPP. Mitsunobu and Related Reactions: Advances and Applications. *Chem. Rev.* **2009**, 109 (6), 2551–2651. (d) Mitsunobu O, Yamada M, and Mukaiyama T. Preparation of Esters of Phosphoric Acid by the Reaction of Trivalent Phosphorus Compounds with Diethyl Azodicarboxylate in the Presence of Alcohols. *Bull. Chem. Soc. Jpn.* **1967**, 40 (4), 935–939. The open access paper can be found at https://doi.org/10.1246/bcsj.40.935 (accessed December 5, 2019). (e) Mitsunobu O, and Yamada M. Preparation of Esters of Carboxylic and Phosphoric Acid via Quaternary Phosphonium Salts. *Bull. Chem. Soc. Jpn.* **1967**, 40 (10), 2380–2382. The open access paper can be found at https://doi.org/10.1246/bcsj.40.2380 (accessed December 5, 2019).

[62] (a) Ishiyama T, Chen H, Morken JP, Mlynarski SN, Ferris GE, Xu S, and Wang J. (2018). 4,4,4′,4′,5,5,5′,5′-Octamethyl-2,2′-bi-1,3,2-dioxaborolane. In e-EROS Encyclopedia of Reagents for Organic Synthesis. doi:10.1002/047084289X.rn00188.pub4. (b) Petasis NA, and Akritopoulou I. The boronic acid mannich reaction: A new method for the synthesis of geometrically pure allylamines. *Tetrahedron Lett.* **1993**, 34 (4), 583–586. (c) Ishiyama T, Murata M, and Miyaura N. Palladium(0)-Catalyzed Cross-Coupling Reaction of Alkoxydiboron with Haloarenes: A Direct Procedure for Arylboronic Esters. *J. Org. Chem.* **1995**, 60 (23), 7508–7510.

[63] (a) Tokuyasu T, Kunikawa S, Masuyama A, and Nojima M. Co(III)–Alkyl Complex- and Co(III)–Alkylperoxo Complex-Catalyzed Triethylsilylperoxidation of Alkenes with Molecular Oxygen and Triethylsilane. *Org. Lett.* **2002**, 4 (21), 3595–3598. (b) Mukaiyama T, Isayama S, Inoki S, Kato K, Yamada T, and Takai T. Oxidation-Reduction Hydration of Olefins with Molecular Oxygen

and 2-Propanol Catalyzed by Bis(acetylacetonato)cobalt(II). *Chem. Lett.* **1989**, 18 (3), 449–452. (c) Inoki S, Kato K, Takai T, Isayama S, Yamada T, and Mukaiyama T. Bis(trifluoroacetylacetonato) cobalt(II) Catalyzed Oxidation-Reduction Hydration of Olefins Selective Formation of Alcohols from Olefins. *Chem. Lett.* **1989**, 18 (3), 515–518. (d) Isayama S, and Mukaiyama T. A New Method for Preparation of Alcohols from Olefins with Molecular Oxygen and Phenylsilane by the Use of Bis(acetylacetonato)cobalt(II). *Chem. Lett.* **1989**, 18 (6), 1071–1074.

[64] (a) Woodward RB, and Hoffmann R. Stereochemistry of Electrocyclic Reactions. *J. Am. Chem. Soc.* **1965**, 87 (2), 395–397. (b) Woodward RB, and Hoffmann R. The Conservation of Orbital Symmetry. *Angew. Chem. Int. Ed. Engl.* **1969**, 8 (11), 781–853. (c) In 1965 Robert Burns Woodward received the Nobel Prize in Chemistry (accessed December 5, 2019, at https://www.nobelprize.org/prizes/chemistry/1965/summary/). In 1981 Roald Hoffmann (jointly with Kenichi Fukui) received the Nobel Prize in Chemistry (accessed December 5, 2019, at https://www.nobelprize.org/prizes/chemistry/1981/summary/). (d) Nazarov IN, and Zaretskaya II. Acetylene derivatives. XVII. Hydration of Hydrocarbons of the Divinylacetylene Series. *Izv. Akad. Nauk. SSSR, Ser. Khim.* **1941**, 211–224. (e) Nazarov IN, and Zaretskaya II. Derivatives of acetylene. XXVII. Hydration of divinylacetylene. *Bull. acad. sci. U.R.S.S., Classe sci. chim.* **1942**, 200–209. Note, the original references are published in Russian and they are difficult to locate. See also (f) Santelli-Rouvier C, and Santelli M. The Nazarov Cyclisation. *Synthesis* **1983** (6), 429–442. (g) Frontier AJ, and Collison C. The Nazarov cyclization in organic synthesis. Recent advances. *Tetrahedron* **2005**, 61 (32), 7577–7606. Other review references can be found in [5c].

[65] (a) Kornblum N, and Brown RA. The Action of Acids on Nitronic Esters and Nitroparaffin Salts. Concerning the Mechanisms of the Nef and the Hydroxamic Acid Forming Reactions of Nitroparaffins. *J. Am. Chem. Soc.* **1965**, 87 (8), 1742–1747. (b) Konovalov MI. *J. Russ. Phys. Chem. Soc.* **1893**, 25, 509. Note, the original reference is published in Russian and it is difficult to locate: Konovalov MI. *Zhur. Russ. Khim. Obshch.* **1893**, 25, 389, 472, 509. (c) Nef JU. Ueber die Constitution der Salze der Nitroparaffine. *Justus Liebigs Ann. Chem.* **1894**, 280 (2–3), 263–291. (d) Nef JU. Ueber das zweiwerthige Kohlenstoffatom. *Justus Liebigs Ann. Chem.* **1894**, 280 (2–3), 291–342.

[66] (a) King AO, Okukado N, and Negishi E. Highly general stereo-, regio-, and chemo-selective synthesis of terminal and internal conjugated enynes by the Pd-catalysed reaction of alkynylzinc reagents with alkenyl halides. *J. Chem. Soc., Chem. Commun.* **1977**, (19), 683–684. (b) Negishi E, King AO, and Okukado N. Selective carbon-carbon bond formation via transition metal catalysis. 3. A highly selective synthesis of unsymmetrical biaryls and diarylmethanes by the nickel- or palladium-catalyzed reaction of aryl- and benzylzinc derivatives with aryl halides. *J. Org. Chem.* **1977**, 42 (10), 1821–1823.

[67] (a) Norrish RGW, and Kirkbride FW. 204. Primary photochemical processes. Part I. The decomposition of formaldehyde. *J. Chem. Soc.* **1932**, (0), 1518–1530. (b) Norrish RGW, and Appleyard MES. 191. Primary photochemical reactions. Part IV. Decomposition of methyl ethyl ketone and methyl butyl ketone. *J. Chem. Soc.* **1934**, (0), 874–880. (c) Norrish RGW, Crone HG, and Saltmarsh OD. 318. Primary photochemical reactions. Part V. The spectroscopy and photochemical decomposition of acetone. *J. Chem. Soc.* **1934**, (0), 1456–1464. (d) Bamford CH, and Norrish RGW. 359. Primary photochemical reactions. Part VII. Photochemical decomposition of *iso*valeraldehyde and di-*n*-propyl ketone. *J. Chem. Soc.* **1935**, (0), 1504–1511. (e) Norrish RGW, and Bamford CH. Photodecomposition of Aldehydes and Ketones. *Nature* **1936**, 138, 1016. (f) Norrish RGW, and Bamford CH. Photo-decomposition of Aldehydes and Ketones. *Nature* **1937**, 140, 195–196. (g) The Nobel Prize in Chemistry 1967. NobelPrize.org. Nobel Media AB 2019 (accessed December 5, 2019, at https://www.nobelprize.org/prizes/chemistry/1967/summary/).

[68] (a) Nelson DJ, Manzini S, Urbina-Blanco CA, and Nolan SP. Key processes in ruthenium-catalysed olefin metathesis. *Chem. Commun.* **2014**, 50 (72), 10355–10375. (b) Ziegler K, Holzkamp E, Breil H, and Martin H. Polymerisation von Äthylen und anderen Olefinen. *Angew. Chem.* **1955**, 67 (16), 426. (c) Anderson AW, and Merckling NG. Polymeric bicyclo-(2,2,1)-2-heptene. US Patent Office 2721189, **1955** (E. I. du Pont de Nemours and Company). (d) The Nobel Prize in Chemistry 2005. NobelPrize.org. Nobel Media AB 2019 (accessed December 5, 2019, at https://www.nobelprize.org/prizes/chemistry/2005/summary/). (e) Takacs JM, and Atkins JM. (2002). Dichloro-bis(tricyclohexylphosphine)methyleneruthenium. In e-EROS Encyclopedia of Reagents for Organic Synthesis, (Ed.). doi:10.1002/047084289X.rn00110. (f) Diver ST, and Middleton MD. (2010). Ruthenium, [1,3-Bis(2,4,6-trimethylphenyl)-2-imidazolidinylidene]dichloro(phenylmethylene)(tricyclohexylphosphine) (Grubbs' Second-Generation Catalyst). In e-EROS Encyclopedia of Reagents for Organic Synthesis, (Ed.). doi:10.1002/047084289X.rn01100. (g) Garber SB, Khan RKM, Mann TJ, and Hoveyda AH. (2013). Dichloro[[2-(1-methylethoxy-*O*)phenyl]-methylene] (tricyclohexylphosphine) Ruthenium. In e-EROS Encyclopedia of Reagents for Organic Synthesis, (Ed.). doi:10.1002/047084289X.rn00129.pub2.

[69] Oppenauer RV. Eine Methode der Dehydrierung von Sekundären Alkoholen zu Ketonen. I. Zur Herstellung von Sterinketonen und Sexualhormonen. *Recl. Trav. Chim. Pays-Bas* **1937**, 56 (2), 137–144.

[70] (a) Criegee R, and Wenner G. Die Ozonisierung des 9,10-Oktalins. *Justus Liebigs Ann. Chem.* **1949**, 564 (1), 9–15. (b) Criegee R. Mechanism of Ozonolysis. *Angew. Chem. Int. Ed. Engl.* **1975**, 14 (11), 745–752. (c) Criegee R. Mechanismus der Ozonolyse. *Angew. Chem.* **1975**, 87 (21), 765–771. (d) Wee AG, Liu B, Jin Z, and Shah AK. (2012). Sodium Periodate–Osmium Tetroxide. In e-EROS Encyclopedia of Reagents for Organic Synthesis, (Ed.). https://onlinelibrary.wiley.com/doi/abs/10.1002/047084289X.rs095m.pub2 (accessed December 5, 2019). (e) Berglund RA, and Kreilein MM. (2006). Ozone. In e-EROS Encyclopedia of Reagents for Organic Synthesis, (Ed.). doi:10.1002/047084289X.ro030.pub2. (f) Wee AG, and Liu B. (2001). Sodium Periodate–Potassium Permanganate. In e-EROS Encyclopedia of Reagents for Organic Synthesis, (Ed.). doi:10.1002/047084289X.rs096. (g) Harries C. Ueber die Einwirkung des Ozons auf organische Verbindungen. *Justus Liebigs Ann. Chem.* **1905**, 343 (2–3), 311–344. Note, the original reference is very old (1840) and it is difficult to locate. See also references in [70g].

[71] (a) Voss J. (2006). 2,4-Bis(4-methoxyphenyl)-1,3,2,4-dithiadiphosphetane 2,4-Disulfide. In e-EROS Encyclopedia of Reagents for Organic Synthesis, (Ed.). doi:10.1002/047084289X.rb170.pub2. (b) Paal C. Ueber die Derivate des Acetophenonacetessigesters und des Acetonylacetessigesters. *Ber. Dtsch. Chem. Ges.* **1884**, 17 (2), 2756–2767. (c) Knorr L. Synthese von Furfuranderivaten aus dem Diacetbernsteinsäureester. *Ber. Dtsch. Chem. Ges.* **1884**, 17 (2), 2863–2870.

[72] (a) Paternò E, and Chieffi G. Sintesi in chimica organica per mezzo della luce. Nota II. Composti degli idrocarburi non saturi con aldeidi e chetoni. *Gazz. Chim. Ital.* **1909**, 39, 341. Note, the original reference is published in Italian and it is difficult to locate. (b) Büchi G, Inman CG, and Lipinsky ES. Light-catalyzed Organic Reactions. I. The Reaction of Carbonyl Compounds with 2-Methyl-2-butene in the Presence of Ultraviolet Light. *J. Am. Chem. Soc.* **1954**, 76 (17), 4327–4331.

[73] (a) Brummonda KM, and Kent JL. Recent Advances in the Pauson–Khand Reaction and Related [2+2+1] Cycloadditions. *Tetrahedron* **2000**, 56 (21), 3263–3283. Please check additional review articles in [5]. (b) Khand IU, Knox GR, Pauson PL, and Watts WE. Organocobalt complexes. Part I. Arene complexes derived from dodecacarbonyltetracobalt. *J. Chem. Soc., Perkin Trans. 1* **1973**, (0), 975–977. (c) Khand IU, Knox GR, Pauson PL, Watts WE, and Foreman MI.

Organocobalt complexes. Part II. Reaction of acetylenehexacarbonyldicobalt complexes, $(R^1C_2R^2)Co_2(CO)_6$, with norbornene and its derivatives. *J. Chem. Soc., Perkin Trans. 1* **1973**, (0), 977–981. (d) Pauson PL, and Khand IU. Uses of Cobalt-Carbonyl Acetylene Complexes in Organic Synthesis. *Ann. N. Y. Acad. Sci.* **1977**, 295 (1), 2–14.

[74] (a) Valeur E, and Bradley M. Amide bond formation: beyond the myth of coupling reagents. *Chem. Soc. Rev.* **2009**, 38 (2), 606–631. (b) El-Faham A, and Albericio F. Peptide Coupling Reagents, More than a Letter Soup. *Chem. Rev.* **2011**, 111 (11), 6557–6602. (c) Carpino LA, Imazumi H, El-Faham A, Ferrer FJ, Zhang C, Lee Y, Foxman BM, Henklein P, Hanay C, Mügge C, Wenschuh H, Klose J, Beyermann M, and Bienert M. The Uronium/Guanidinium Peptide Coupling Reagents: Finally the True Uronium Salts. *Angew. Chem. Int. Ed.* **2002**, 41 (3), 441–445. (d) Albert JS, Hamilton AD, Hart AC, Feng X, Lin L, and Wang Z. (2017). 1,3-Dicyclohexylcarbodiimide. In e-EROS Encyclopedia of Reagents for Organic Synthesis. doi:10.1002/047084289X.rd146.pub3. (e) Pottorf RS, Szeto P, and Srinivasarao M. (2017). 1-Ethyl-3-(3'-dimethylaminopropyl)carbodiimide Hydrochloride. In e-EROS Encyclopedia of Reagents for Organic Synthesis. doi:10.1002/047084289X.re062.pub2. (f) Albericio F, and Kates SA. (2001). *O*-Benzotriazol-1-yl-*N*,*N*,*N'*,*N'*-tetramethyluronium Hexafluorophosphate. In e-EROS Encyclopedia of Reagents for Organic Synthesis, (Ed.). doi:10.1002/047084289X.rb038. (g) Albericio F, Kates SA, and Carpino LA. (2001). *N*-[(Dimethylamino)-1*H*-1,2,3-triazolo[4,5-*b*]pyridin-1-ylmethylene]-*N*-methylmethanaminium Hexafluorophosphate *N*-Oxide. In e-EROS Encyclopedia of Reagents for Organic Synthesis, (Ed.). https://onlinelibrary.wiley.com/doi/10.1002/047084289X.rd312s (accessed December 5, 2019). (h) Coste J, and Jouin P. (2003). (1*H*-Benzotriazol-1-yloxy)tripyrrolidino-phosphonium Hexafluorophosphate. In e-EROS Encyclopedia of Reagents for Organic Synthesis, (Ed.). doi:10.1002/047084289X.rn00198. (i) Lygo B, and Pelletier G. (2013). 1-Hydroxybenzotriazole. In e-EROS Encyclopedia of Reagents for Organic Synthesis, (Ed.). doi:10.1002/047084289X.rh052.pub2. (j) Fischer E, and Fourneau E. Ueber einige Derivate des Glykocolls. *Ber. Dtsch. Chem. Ges.* **1901**, 34 (2), 2868–2877. (k) Sheehan JC, and Hess GP. A New Method of Forming Peptide Bonds. *J. Am. Chem. Soc.* **1955**, 77 (4), 1067–1068. (l) Dourtoglou V, Ziegler JC, and Gross B. L'hexafluorophosphate de O-benzotriazolyl-N,N-tetramethyluronium: Un reactif de couplage peptidique nouveau et efficace. *Tetrahedron Lett.* **1978**, 19 (15), 1269–1272.

[75] (a) Baldwin JE. Rules for ring closure. *J. Chem. Soc., Chem. Commun.* **1976**, (18), 734–736. (b) Pictet A, and Spengler T. Über die Bildung von Isochinolin-derivaten durch Einwirkung von Methylal auf Phenyl-äthylamin, Phenyl-alanin und Tyrosin. *Ber. Dtsch. Chem. Ges.* **1911**, 44 (3), 2030–2036.

[76] (a) Demjanov NJ, and Lushnikov M. *J. Russ. Phys. Chem. Soc.* **1903**, 35, 26–42. Note, the original reference is published in Russian and it is difficult to locate: Demjanov NJ, and Lushnikov M. *Zhur. Russ. Khim. Obshch.* **1903**, 35, 26–42. (b) Tiffeneau M, Weill P, and Tchoubar B. Isomérisation de l'oxyde de méthylène cyclohexane en hexahydrobenzaldéhyde et désamination de l'aminoalcool correspondant en cycloheptanone. *Compt. Rend.* **1937**, 205, 54–56. The reference is in French and can be viewed at https://gallica.bnf.fr/ark:/12148/bpt6k3157c.image.f54 (accessed December 5, 2019). (c) Fittig R. 41. Ueber einige Derivate des Acetons. *Justus Liebigs Ann. Chem.* **1860**, 114 (1), 54–63.

[77] (a) Cave A, Kan-Fan C, Potier P, and Men JL. Modification de la reaction de polonovsky: Action de l'anhydride trifluoroacetique sur un aminoxide. *Tetrahedron* **1967**, 23 (12), 4681–4689. (b) Ahond A, Cave A, Kan-Fan C, Husson HP, Rostolan J, and Potier P. Facile N-O bond cleavages of amine oxides. *J. Am. Chem. Soc.* **1968**, 90 (20), 5622–5623. (c) Polonovski M, and Polonovski M. Sur les aminoxydes des alcaloides. III. Action des anhydrides et chlorules d'acides organiques. Preparations des bases nor. *Bull. Soc. Chim. Fr.* **1927**, 41, 1190–1208. Note, the original reference is published in French and it is difficult to locate.

[78] (a) Katsuki T, and Sharpless KB. The first practical method for asymmetric epoxidation. *J. Am. Chem. Soc.* **1980**, 102 (18), 5974–5976. (b) Tu Y, Wang ZX, and Shi Y. An Efficient Asymmetric Epoxidation Method for *trans*-Olefins Mediated by a Fructose-Derived Ketone. *J. Am. Chem. Soc.* **1996**, 118 (40), 9806–9807. (c) Prileschajew N. Oxydation ungesättigter Verbindungen mittels organischer Superoxyde. *Ber. Dtsch. Chem. Ges.* **1909**, 42 (4), 4811–4815.

[79] (a) Dobbs AP, Guesné SJJ, Parker RJ, Skidmore J, Stephensond RA, and Hursthouse MB. A detailed investigation of the aza-Prins reaction. *Org. Biomol. Chem.* **2010**, 8 (5), 1064–1080. (b) Reddy BVS, Nair PN, Antony A, Lalli C, and Grée R. The Aza-Prins Reaction in the Synthesis of Natural Products and Analogues. *Eur. J. Org. Chem.* **2017**, (14), 1805–1819. (c) Rajasekaran P, Singh GP, Hassam M, and Vankar YD. A Cascade "Prins-Pinacol-Type Rearrangement and C4-OBn Participation" on Carbohydrate Substrates: Synthesis of Bridged Tricyclic Ketals, Annulated Sugars and C2-Branched Heptoses. *Chem. Eur. J.* **2016**, 22 (51), 18383–18387. (d) Prins HJ. Over de condensatie van formaldehyd met onverzadigde verbindingen. *Chem. Weekblad* **1919**, 16, 1072–1073. (e) Prins HJ. The reciprocal condensation of unsaturated organic compounds. *Chem. Weekblad* **1919**, 16, 1510–1526. Note, the original references are difficult to locate.

[80] Pummerer R. Über Brom-Additionsprodukte von Aryl-thioglykolsäuren. *Ber. Dtsch. Chem. Ges.* **1909**, 42 (2), 2275–2282.

[81] (a) See more at https://en.wikipedia.org/wiki/Cheletropic_reaction (accessed December 5, 2019). (b) Philips JC, and Morales O. Sulphur dioxide extrusion from substituted thiiren 1,1-dioxides. *J. Chem. Soc., Chem. Commun.* **1977**, (20), 713–714. (c) Ramberg L, and Bäcklund B. *Ark. Kemi. Mineral. Geol.* **1940**, 27, Band 13A, 1–50. Note, the original reference is difficult to locate (see also *Chem. Abstr.* **1940**, 34, 4725).

[82] (a) Blaise EE. *Comp. Rend. Hebd. Seances Acad. Sci.* **1901**, 132, 478–480. Note, the original reference is difficult to locate. (b) Cason J, Rinehart KL, and Thornton SD. The Preparation of β-Keto Esters from Nitriles and α-Bromoesters. *J. Org. Chem.* **1953**, 18 (11), 1594–1600. (c) Reformatsky S. Neue Synthese zweiatomiger einbasischer Säuren aus den Ketonen. *Ber. Dtsch. Chem. Ges.* **1887**, 20 (1), 1210–1211.

[83] (a) Rapson WS, and Robinson R. 307. Experiments on the synthesis of substances related to the sterols. Part II. A new general method for the synthesis of substituted cyclohexenones. *J. Chem. Soc.* **1935**, (0), 1285–1288. (b) The Nobel Prize in Chemistry 1947. NobelPrize.org. Nobel Media AB 2019 (accessed December 5, 2019, at https://www.nobelprize.org/prizes/chemistry/1947/summary/).

[84] (a) Bamford WR, and Stevens TS. 924. The decomposition of toluene-*p*-sulphonylhydrazones by alkali. *J. Chem. Soc.* **1952**, (0), 4735–4740. (b) Shapiro RH, and Heath MJ. Tosylhydrazones. V. Reaction of Tosylhydrazones with Alkyllithium Reagents. A New Olefin Synthesis. *J. Am. Chem. Soc.* **1967**, 89 (22), 5734–5735. (c) Shapiro RH, Lipton MF, Kolonko KJ, Buswell RL, and Capuano LA. Tosylhydrazones and alkyllithium reagents: More on the regiospecificity of the reaction and the trapping of three intermediates. *Tetrahedron Lett.* **1975**, 16 (22–23), 1811–1814. (d) Shapiro RH. (2011). Alkenes from Tosylhydrazones. In Organic Reactions, (Ed.). doi:10.1002/0471264180.or023.03.

[85] (a) Stephens RD, and Castro CE. The Substitution of Aryl Iodides with Cuprous Acetylides. A Synthesis of Tolanes and Heterocyclics. *J. Org. Chem.* **1963**, 28 (12), 3313–3315. (b) Sonogashira K, Tohda Y, and Hagihara N. A convenient synthesis of acetylenes: catalytic substitutions of acetylenic hydrogen with bromoalkenes, iodoarenes and bromopyridines. *Tetrahedron Lett.* **1975**, 16 (50), 4467–4470.

[86] (a) Staudinger H. Zur Kenntniss der Ketene. Diphenylketen. *Justus Liebigs Ann. Chem.* **1907**, 356 (1–2), 51–123. (b) Staudinger H. Über Ketene. 4. Mitteilung: Reaktionen des Diphenylketens. *Ber. Dtsch. Chem. Ges.* **1907**, 40 (1), 1145–1148. (c) Saxon E, and Bertozzi CR. Cell Surface

Engineering by a Modified Staudinger Reaction. *Science* **2000**, 287 (5460), 2007–2010. (d) Saxon E, Armstrong JI, and Bertozzi CR. A "Traceless" Staudinger Ligation for the Chemoselective Synthesis of Amide Bonds. *Org. Lett.* **2000**, 2 (14), 2141–2143. (e) Wang ZPA, Tiana CL, and Zheng JS. The recent developments and applications of the traceless-Staudinger reaction in chemical biology study. *RSC Adv.* **2015**, 5 (130), 107192–107199. (f) Staudinger H, and Meyer J. Über neue organische Phosphorverbindungen III. Phosphinmethylenderivate und Phosphinimine. *Helv. Chim. Acta* **1919**, 2 (1), 635–646. (g) The Nobel Prize in Chemistry 1953. NobelPrize.org. Nobel Media AB 2019 (accessed December 5, 2019, at https://www.nobelprize.org/prizes/chemistry/1953/summary/).

[87] (a) Albert JS, and Hamilton AD. (2001). 1,3-Dicyclohexylcarbodiimide–4-Dimethylaminopyridine. In e-EROS Encyclopedia of Reagents for Organic Synthesis, (Ed.). doi:10.1002/047084289X.rd147. (b) Neises B, and Steglich W. Simple Method for the Esterification of Carboxylic Acids. *Angew. Chem. Int. Ed. Engl.* **1978**, 17 (7), 522–524.

[88] (a) Merrifield JH, Godschalx JP, and Stille JK. Synthesis of unsymmetrical diallyl ketones: the palladium-catalyzed coupling of allyl halides with allyltin reagents in the presence of carbon monoxide. *Organometallics* **1984**, 3 (7), 1108–1112. (b) Tokuyama H, Yokoshima S, Yamashita T, and Fukuyama T. A novel ketone synthesis by a palladium-catalyzed reaction of thiol esters and organozinc reagents. *Tetrahedron Lett.* **1998**, 39 (20), 3189–3192. (c) Milstein D, and Stille JK. A general, selective, and facile method for ketone synthesis from acid chlorides and organotin compounds catalyzed by palladium. *J. Am. Chem. Soc.* **1978**, 100 (11), 3636–3638. (d) Milstein D, and Stille JK. Palladium-catalyzed coupling of tetraorganotin compounds with aryl and benzyl halides. Synthetic utility and mechanism. *J. Am. Chem. Soc.* **1979**, 101 (17), 4992–4998.

[89] (a) Carrow BP, and Hartwig JF. Distinguishing Between Pathways for Transmetalation in Suzuki–Miyaura Reactions. *J. Am. Chem. Soc.* **2011**, 133 (7), 2116–2119. (b) Thomas AA, and Denmark SE. Pre-transmetalation intermediates in the Suzuki-Miyaura reaction revealed: The missing link. *Science* **2016**, 352 (6283), 329–332. (c) Miyaura N, Yamada K, and Suzuki A. A new stereospecific cross-coupling by the palladium-catalyzed reaction of 1-alkenylboranes with 1-alkenyl or 1-alkynyl halides. *Tetrahedron Lett.* **1979**, 20 (36), 3437–3440. (d) Miyaura N, and Suzuki A. Stereoselective synthesis of arylated (*E*)-alkenes by the reaction of alk-1-enylboranes with aryl halides in the presence of palladium catalyst. *J. Chem. Soc., Chem. Commun.* **1979**, (19), 866–867.

[90] (a) Omura K, Sharma AK, and Swern D. Dimethyl sulfoxide-trifluoroacetic anhydride. New reagent for oxidation of alcohols to carbonyls. *J. Org. Chem.* **1976**, 41 (6), 957–962. (b) Tidwell TT. (2001). Dimethyl Sulfoxide–Dicyclohexylcarbodiimide. In e-EROS Encyclopedia of Reagents for Organic Synthesis, (Ed.). doi:10.1002/047084289X.rd375. Please also see (c) Omura K, and Swern D. Oxidation of alcohols by "activated" dimethyl sulfoxide. A preparative, steric and mechanistic study. *Tetrahedron* **1978**, 34 (11), 1651–1660. (d) Mancuso AJ, Brownfain DS, and Swern D. Structure of the dimethyl sulfoxide-oxalyl chloride reaction product. Oxidation of heteroaromatic and diverse alcohols to carbonyl compounds. *J. Org. Chem.* **1979**, 44 (23), 4148–4150.

[91] (a) Passerini M, and Simone L. Sopra gli isonitrili (I). Composto del *p*-isonitril-azobenzolo con acetone ed acido acetico. *Gazz. Chim. Ital.* **1921**, 51, 126–129. (b) Passerini M. Sopra gli isonitrili (II). Composti con aldeidi o con chetoni ed acidi organici monobasici. *Gazz. Chim. Ital.* **1921**, 51, 181–189. Note, the original references are published in Italian and they are difficult to locate. (c) Ugi I, Meyr R, and Fetzer U. Versammlungsberichte. *Angew. Chem.* **1959**, 71 (11), 373–388. The citation can be found at https://onlinelibrary.wiley.com/doi/abs/10.1002/ange.19590711110 (accessed December 5, 2019).

[92] (a) Ullmann F, and Bielecki J. Ueber Synthesen in der Biphenylreihe. *Ber. Dtsch. Chem. Ges.* **1901**, 34 (2), 2174–2185. (b) Ullmann F. Ueber symmetrische Biphenylderivate. *Justus*

Liebigs Ann. Chem. **1904**, 332 (1–2), 38–81. (c) Ullmann F. Ueber eine neue Bildungsweise von Diphenylaminderivaten. *Ber. Dtsch. Chem. Ges.* **1903**, 36 (2), 2382–2384. (d) Ullmann F, and Sponagel P. Ueber die Phenylirung von Phenolen. *Ber. Dtsch. Chem. Ges.* **1905**, 38 (2), 2211–2212. (e) Sambiagio C, Marsden SP, Blackera AJ, and McGowan PC. Copper catalysed Ullmann type chemistry: from mechanistic aspects to modern development. *Chem. Soc. Rev.* **2014**, 43 (10), 3525–3550.

[93] (a) Gao Y, and Cheun Y. (2013). Osmium Tetroxide–*N*-Methylmorpholine *N*-Oxide. In e-EROS Encyclopedia of Reagents for Organic Synthesis, (Ed.). doi:10.1002/047084289X.ro009.pub2. (b) Hentges SG, and Sharpless KB. Asymmetric induction in the reaction of osmium tetroxide with olefins. *J. Am. Chem. Soc.* **1980**, 102 (12), 4263–4265. (c) Wai JSM, Marko I, Svendsen JS, Finn MG, Jacobsen EN, and Sharpless KB. A mechanistic insight leads to a greatly improved osmium-catalyzed asymmetric dihydroxylation process. *J. Am. Chem. Soc.* **1989**, 111 (3), 1123–1125. (d) Ogino Y, Chen H, Kwong HL, and Sharpless KB. On the timing of hydrolysis / reoxidation in the osmium-catalyzed asymmetric dihydroxylation of olefins using potassium ferricyanide as the reoxidant. *Tetrahedron Lett.* **1991**, 32 (32), 3965–3968. (e) Lee DG, Ribagorda M, and Adrio J. (2007). Potassium Permanganate. In e-EROS Encyclopedia of Reagents for Organic Synthesis, (Ed.). doi:10.1002/9780470842898.rp244.pub2. (f) VanRheenen V, Kelly RC, and Cha DY. An improved catalytic OsO_4 oxidation of olefins to cis-1,2-glycols using tertiary amine oxides as the oxidant. *Tetrahedron Lett.* **1976**, 17 (23), 1973–1976. (g) The Nobel Prize in Chemistry 2001. NobelPrize.org. Nobel Media AB 2019 (accessed December 5, 2019, at https://www.nobelprize.org/prizes/chemistry/2001/summary/).

[94] (a) Reimer K, and Tiemann F. Ueber die Einwirkung von Chloroform auf alkalische Phenolate. *Ber. Dtsch. Chem. Ges.* **1876**, 9 (1), 824–828. (b) Vilsmeier A, and Haack A. Über die Einwirkung von Halogenphosphor auf Alkyl-formanilide. Eine neue Methode zur Darstellung sekundärer und tertiärer *p*-Alkylamino-benzaldehyde. *Ber. Dtsch. Chem. Ges.* A/B **1927**, 60 (1), 119–122.

[95] (a) Keith JA, Nielsen RJ, Oxgaard J, and Goddard WA. Unraveling the Wacker Oxidation Mechanisms. *J. Am. Chem. Soc.* **2007**, 129 (41), 12342–12343. (b) Keith JA, and Henry PM. The Mechanism of the Wacker Reaction: A Tale of Two Hydroxypalladations. *Angew. Chem. Int. Ed.* **2009**, 48 (48), 9038–9049. (c) Smidt J, Hafner W, Jira R, Sedlmeier J, Sieber R, Rüttinger R, and Kojer H. Katalytische Umsetzungen von Olefinen an Platinmetall-Verbindungen Das Consortium-Verfahren zur Herstellung von Acetaldehyd. *Angew. Chem.* **1959**, 71 (5), 176–182.

[96] (a) Wagner G. *J. Russ. Phys. Chem. Soc.* **1899**, 31, 690–693. Note, the original reference is published in Russian and it is difficult to locate. (b) Wagner G, and Brickner W. Ueber die Beziehung der Pinenhaloïdhydrate zu den Haloïdanhydriden des Borneols. *Ber. Dtsch. Chem. Ges.* **1899**, 32 (2), 2302–2325. (c) Meerwein H. Über den Reaktionsmechanismus der Umwandlung von Borneol in Camphen; [Dritte Mitteilung über Pinakolinumlagerungen.]. *Justus Liebigs Ann. Chem.* **1914**, 405 (2), 129–175.

[97] (a) Weinreb SM, Folmer JJ, Ward TR, and Georg GI. (2006). *N,O*-Dimethylhydroxylamine. In e-EROS Encyclopedia of Reagents for Organic Synthesis, (Ed.). https://onlinelibrary.wiley.com/doi/10.1002/047084289X.rd341.pub2 (accessed December 5, 2019). (b) Hisler K, Tripoli R, and Murphy JA. Reactions of Weinreb amides: formation of aldehydes by Wittig reactions. *Tetrahedron Lett.* **2006**, 47 (35), 6293–6295. (c) Nahm S, and Weinreb SM. N-methoxy-N-methylamides as effective acylating agents. *Tetrahedron Lett.* **1981**, 22 (39), 3815–3818.

[98] (a) Byrnea PA, and Gilheany DG. The modern interpretation of the Wittig reaction mechanism. *Chem. Soc. Rev.* **2013**, 42 (16), 6670–6696. (b) Schlosser M, and Christmann KF. Trans-Selective Olefin Syntheses. *Angew. Chem. Int. Ed. Engl.* **1966**, 5 (1), 126–126. (c) Wittig G, and Schöllkopf U. Über Triphenyl-phosphin-methylene als olefinbildende Reagenzien (I. Mitteil.). *Chem. Ber.* **1954**, 87 (9), 1318–1330. (d) Wittig G, and Haag W. Über Triphenyl-phosphinmethylene als olefinbildende Reagenzien (II. Mitteil.). *Chem. Ber.* **1955**, 88 (11), 1654–1666.

[99] (a) Wohl A. Bromierung ungesättigter Verbindungen mit *N*-Brom-acetamid, ein Beitrag zur Lehre vom Verlauf chemischer Vorgänge. *Ber. Dtsch. Chem. Ges.* A/B **1919**, 52 (1), 51–63. (b) Ziegler K, Späth A, Schaaf E, Schumann W, and Winkelmann E. Die Halogenierung ungesättigter Substanzen in der Allylstellung. *Justus Liebigs Ann. Chem.* **1942**, 551 (1), 80–119. (c) The Nobel Prize in Chemistry 1963. NobelPrize.org. Nobel Media AB 2019 (accessed December 5, 2019, at https://www.nobelprize.org/prizes/chemistry/1963/summary/).

[100] (a) Huang-Minlon. A Simple Modification of the Wolff-Kishner Reduction. *J. Am. Chem. Soc.* **1946**, 68 (12), 2487–2488. (b) Clemmensen E. Reduktion von Ketonen und Aldehyden zu den entsprechenden Kohlenwasserstoffen unter Anwendung von amalgamiertem Zink und Salzsäure. *Ber. Dtsch. Chem. Ges.* **1913**, 46 (2), 1837–1843. (c) Kishner N. *J. Russ. Phys. Chem. Soc.* **1911**, 43, 582–592. Note, the original reference is published in Russian and it is difficult to locate. (d) Wolff L. Chemischen Institut der Universität Jena: Methode zum Ersatz des Sauerstoffatoms der Ketone und Aldehyde durch Wasserstoff. [Erste Abhandlung.]. *Justus Liebigs Ann. Chem.* **1912**, 394 (1), 86–108.

中文索引

A

阿恩特 – 艾斯特尔特合成 / 024

阿尔布佐夫反应 / 022

安息香缩合 / 034

B

巴顿脱羧 / 028

拜耳 - 维利格氧化 / 026

保森 - 坎德反应 / 156

贝克曼重排 / 032

贝里斯 - 希尔曼反应 / 030

苯炔机理 / 036

比施勒 - 纳皮耶拉尔斯基环化 / 042

波罗诺夫斯基反应 / 166

伯格曼环化 / 038

伯奇还原 / 040

布赫瓦尔德 - 哈特维希交叉偶联 / 046

布朗硼氢化 / 044

C

陈 - 埃文斯 - 兰交叉偶联 / 050

臭氧分解 / 148

D

达森缩合 / 070

戴斯 - 马丁氧化 / 072

狄尔斯 - 阿尔德环加成 / 076

E

厄普约翰双羟基化反应 / 198

F

法沃尔斯基重排 / 080

芳香亲电取代机理 / 006

芳香亲核取代机理 / 008

芳香自由基亲核取代机理 / 010

费歇尔吲哚合成 / 082

傅 - 克酰基化和傅 - 克烷基化 / 084

G

盖布瑞尔合成 / 086

格拉泽 - 埃格林顿 - 海偶联 / 090

格罗布裂解 / 094

格氏反应 / 092

格瓦尔德反应 / 088

根岸交叉偶联 / 138

宫浦硼化 / 130

光延反应 / 128

桧山交叉偶联 / 102

H

赫克交叉偶联 / 098

黑尔 - 福尔哈德 - 泽林斯基反应 / 100

霍夫曼消除 / 104

霍纳 - 沃兹沃斯 - 埃蒙斯烯烃化 / 106

K

坎尼扎罗反应 / 048

柯普消除 / 058

柯普重排 / 060

柯提斯重排 / 066

克莱森缩合 / 054

克莱森重排 / 056

克里奇氧化和马拉普拉德氧化 / 062

库切罗夫反应 / 110

L

莱伊 - 格里菲斯氧化 / 114

兰伯格 - 贝克伦德重排 / 174

雷福尔马茨基反应 / 176

利贝斯金德 - 斯罗格尔交叉偶联 / 116

铃木交叉偶联反应 / 190

卤仿反应 / 096

罗宾逊环化 / 178

M

迈克尔加成 / 124

麦尔外因 - 庞多夫 - 维利还原 / 122

麦克默里偶联 / 120

曼尼希反应 / 118

米尼希反应 / 126

N

纳扎罗夫环化 / 134

尼夫反应 / 136

诺里什 I 型和 II 型反应 / 140

O

欧芬脑尔氧化 / 146

P

帕尔 - 克诺尔合成 / 150

帕特罗 - 布奇反应 / 154

皮克特 - 斯彭格勒反应 / 162

频哪醇 - 频哪酮重排 / 164

普里莱扎耶夫环氧化 / 168

普林斯反应 / 170

普默勒重排 / 172

Q

齐齐巴宾氨基化 / 052

亲电加成机理 / 002

亲核取代机理 / 004

琼斯氧化 / 108

炔烃拉链反应 / 020

S

施蒂勒交叉偶联反应 / 188

施陶丁格反应 / 184

双 -π- 甲烷重排 / 078

斯特格里奇酯化 / 186

斯文氧化 / 192

T

肽（酰胺）偶联 / 158

铜催化叠氮 - 炔环加成 / 064

酮醇缩合 / 018

W

瓦格纳 - 麦尔外因重排 / 204

瓦克氧化 / 202

维蒂希反应 / 208

维尔斯迈尔 - 哈克反应 / 200

温勒伯酮合成 / 206

沃尔夫 - 凯惜纳还原 / 212

沃尔 - 齐格勒反应 / 210

乌尔曼芳基 - 芳基偶联 / 196

乌吉反应 / 194

烯烃复分解 / 142

X

夏皮罗反应 / 180

向山氧化还原水合 / 132

消除机理 / 014

熊田交叉偶联 / 112

Z

菌头交叉偶联反应 / 182

重氮化反应 / 074

英文索引

A

acyloin condensation / 018

alkyne zipper reaction / 020

Arbuzov reaction / 022

Arndt-Eistert synthesis / 024

aromatic electrophilic substitution mechanism / 006

aromatic nucleophilic substitution mechanism / 008

aromatic radical nucleophilic substitution mechanism / 010

B

Baeyer-Villiger oxidation / 026

Barton decarboxylation / 028

Baylis–Hillman reaction / 030

Beckmann rearrangement / 032

benzoin condensation / 034

benzyne mechanism / 036

Bergman cyclization / 038

Birch reduction / 040

Bischler-Napieralski cyclization / 042

Brown hydroboration / 044

Buchwald-Hartwig cross coupling / 046

C

Cannizzaro reaction / 048

Chan-Evans-Lam cross coupling / 050

Chichibabin amination / 052

Claisen condensation / 054

Claisen rearrangement / 056

Cope elimination / 058

Cope rearrangement / 060

copper-catalyzed azide-alkyne cycloaddition, / 064

Criegee & Malaprade oxidation / 062

CuAAC / 064

Curtius rearrangement / 066

D

Darzens condensation / 070

Dess-Martin oxidation / 072

diazotization / 074

Diels-Alder cycloaddition / 076

di-π-methane rearrangement / 078

E

electrophilic addition mechanism / 002

elimination mechanism / 014

F

Favorskii rearrangement / 080

Fischer indole synthesis / 082

Friedel-Crafts acylation & alkylation / 084

G

Gabriel synthesis / 086

Gewald reaction / 088

Glaser-Eglinton-Hay coupling / 090

Grignard reaction / 092

Grob fragmentation / 094

H

haloform reaction / 096

Heck cross coupling / 098

Hell-Volhard-Zelinsky reaction / 100

Hiyama cross coupling / 102

Hofmann elimination / 104

Horner-Wadsworth-Emmons olefination / 106

J

Jones oxidation / 108

K

Kucherov reaction / 110

Kumada cross coupling / 112

L

Ley–Griffith oxidation / 114

Liebeskind-Srogl cross coupling / 116

M

Mannich reaction / 118

McMurry coupling / 120

Meerwein-Ponndorf-Verley reduction / 122

Michael addition / 124

Minisci reaction / 126

Mitsunobu reaction / 128

Miyaura borylation / 130

Mukaiyama RedOx hydration / 132

N

Nazarov cyclization / 134

Nef reaction / 136

Negishi cross coupling / 138

Norrish type I & II reaction / 140

nucleophilic substitution mechanism / 004

O

oefin alkene metathesis / 142

Oppenauer oxidation / 146

ozonolysis / 148

P

Paal–Knorr syntheses / 150

Paternò-Büchi reaction / 154

Pauson-Khand reaction / 156

peptide amide coupling / 158

Pictet–Spengler reaction / 162

pinacol-pinacolone rearrangement / 164

Polonovski reaction / 166

Prilezhaev epoxidation / 168

Prins reaction / 170

Pummerer rearrangement / 172

R

Ramberg-Bäcklund rearrangement / 174

Reformatsky reaction / 176

Robinson annulation / 178

S

Shapiro reaction / 180

Sonogashira cross coupling / 182

Staudinger reaction / 184

Steglich esterification / 186

Stille cross coupling / 188

Suzuki cross coupling / 190

Swern oxidation / 192

U

Ugi reaction / 194

Ullmann aryl-aryl coupling / 196

Upjohn dihydroxylation / 198

V

Vilsmeier-Haack reaction / 200

W

Wacker oxidation / 202

Wagner-Meerwein rearrangement / 204

Weinreb ketone synthesis / 206

Wittig reaction / 208

Wohl-Ziegler reaction / 210

Wolff-Kishner reduction / 212